The B
Boo

How hi-fi works & how to get the best from it

Dan Everard

A WOODHEAD-FAULKNER PUBLICATION

Woodhead-Faulkner Ltd
8 Market Passage
Cambridge CB2 3PF

First published 1977

© Bang & Olufsen UK Limited/Dan Everard 1977

ISBN 0 85941 065 X

Conditions of sale
All rights reserved. No part of this publication may be reproduced, stored in a retrieval system or transmitted, in any form or by any means, electronic, mechanical, photocopying, recording or otherwise, without the prior permission of the copyright holders.

Production services by Book Production Consultants

Printed in Great Britain by
Lowe & Brydone Printers Ltd, Thetford, Norfolk

PREFACE

Over the last few years people have asked my advice on high-fidelity equipment, and it struck me that many of them simply had nowhere to look for basic facts. In the light of this it seemed that spending a lot of money should ensure good equipment, but the results didn't seem to justify the expenditure. Sadly the reasons were simple but expensive to correct since correction meant changing equipment, so often matters were left where they were at second best – but still expensive. The result of this was that I found myself saying "someone ought to write a book".

Here it is.

Dan Everard

In sponsoring this publication, Bang & Olufsen have endeavoured to ensure that the contents in no way unfairly promote their own product philosophy.

This book is based on the generally accepted principles of high fidelity sound reproduction and presents the freely expressed views of the author without regard to the commercial considerations of Bang & Olufsen.

CONTENTS

		Page
Preface		3
Introduction		7
1.	Choosing a Hi-fi System Single unit or discrete system? Frequency response. Distortion. Stereo. Quadrophony and ambiophony. Compatibility. Test records.	9
2.	The Elements of a Hi-fi System	18
3.	The Record and the Record Player The record. The cartridge. The stylus. The pick-up arm. The turntable. Setting up.	21
4.	The Radio Tuner Stereo radio. AM or FM? Aerials.	37
5.	The Amplifier RIAA equalisation. Input selector switch. Tone controls and filters. Volume and loudness controls. Balance control. Stereo/mono switch. The power amplifier.	48
6.	The Loudspeaker Resonances. Drivers and crossover circuits. Elliptical speakers. Loudspeaker cabinets.	57

Contents

Page

Phase linearity. General considerations. Headphones.

7. The Tape Recorder — 70
Types of tape machine. Operation of a tape recorder. Track width and tape speed considerations. Tape drive mechanisms. Metering. Tape recorder controls. Tapes. Dolby system.

8. Setting up a Hi-fi System — 86
Surfaces. Earth loops. Loudspeaker phasing. Positioning in the room.

9. Maintaining a Hi-fi System — 92
General. Record players. Records. Tape recorders. Tapes.

10. The Nature of Sound — 96
Cause and propagation. Energy and hearing sensitivity. Waveforms. Distortion. Echo. Reverberation.

11. Recording — 106
Tape recorder levels. Microphones. Editing.

Glossary — 114

INTRODUCTION

High fidelity (of radio receiver etc.): reproducing sound faithfully (colloq. abbr. *hi-fi*) – *The Concise Oxford Dictionary*

The purpose of this book is to provide the reader with enough knowledge to make basic decisions in choosing a hi-fi system. It is not intended to be a definitive book on the state of the art today, but most of the modern trends which affect the beginner's choices are covered. The descriptions of equipment and reasons for points of design are made as easy to follow as possible. I have tried not to assume any great technical knowledge. In places I am aware that the descriptions are not rigorous; my intention is to give reasons for things, not analyses of how technical problems are solved.

The assumption is made that the prospective hi-fi system is to be stereo. Part of my intention is to provide some understanding of the jargon used in hi-fi descriptions. If a word is not defined in the text or glossary of this book, it does not mean that it does not exist, but do please get someone to define it for you so that you understand it if it affects your final choice of equipment.

Hi-fi equipment is very expensive, so decide how much money you wish to spend on how large a system, *i.e.* whether it is to be a basic record player, amplifier and a

Introduction

pair of speakers, or whether you want a tape recorder and/or radio tuner as well. In Chapter 1 there is a guide on how the total cost should be split between the various parts. Once having decided on a total budget, do not be talked into paying more for something you do not want.

When you know approximately how much you have to spend on each part of the system, go and look at what is available in that price bracket, bearing in mind the points outlined in the relevant chapters. Then, if possible, listen to the likely-looking pieces of equipment; it must sound right to you – after all it is going to be yours.

I would recommend that you read the whole book before committing yourself to buying equipment. This will give you a broad grounding in the subject which will probably help you in coming to a final decision.

1 | CHOOSING A HI-FI SYSTEM

SINGLE UNIT OR DISCRETE SYSTEM?

The choice of a hi-fi system depends not only on its technical performance but also on domestic considerations. The room in which it is to be used will usually be a living-room as well, and the equipment should not overpower it. Obviously it depends how important listening to music is in one's life as to how dominant the equipment is in a room. In choosing the parts of a system this should be borne in mind as well as the technical points outlined in later chapters.

A choice has to be made right at the start: should one buy a single unit containing all the parts and often built as a piece of furniture, or buy the elements as separate units to be wired together and frequently placed in different parts of the room – record player and amplifier in one place, speakers in another – as convenience dictates?

If a single unit is bought, it is usually impossible to improve it at a later date as one's ears become used to good-quality reproduction, and it would normally have to be sold and another better one bought in order to improve the quality. One's hearing improves and becomes more sensitive through practice and, if one is interested in eventually having the best rather than just having something good to play records on, complete replacement of a system can be a heavy expense. Wiring together a

Choosing a Hi-Fi System

discrete system is not difficult as the cables can all be bought already made up and it is simply a case of plugging them in according to the instructions. A discrete system also enables a compromise to be made at the outset between ready cash and quality, allowing improvements to be made piece by piece as they can be afforded. The rest of this book is written as if a discrete system is being chosen, but the statements apply in general to both types.

RELATIVE COSTS

As with most things in life, you only get what you pay for, so equipment costing £100 will not be as good as that costing £150. Their specifications may be identical, but these do not necessarily tell the whole story and, despite strenuous efforts by most people involved, there are still discrepancies in the way parameters are measured. Also there is no way of specifying reliability, and in more expensive equipment this is part of what you pay for.

As a guide to the relative values of record-playing equipment, the costs should be roughly these proportions of the total:

30 per cent – record deck and cartridge (too cheap a deck increases the risk of damaged records).
20 per cent – amplifier.
25 per cent – each speaker.

A tape recorder should be as good as you can afford. Cheap ones are like cheap cameras, good for snapshots; expensive ones will give an accurate copy of what was recorded; very expensive ones do so for longer if they are looked after properly – the instructions will tell you how.

A radio tuner is much the same except that if the radio signal strength in your area is weak or very strong, you will probably need a more expensive one. In the first case this is necessary for additional sensitivity; in the second to

Choosing a Hi-Fi System

be able to withstand an overloaded input without additional distortion or pick-up of unwanted sound. The rest of the choice depends mostly on what facilities the tuner has to make it easier to use or more versatile. Bear in mind that an aerial may be required. A good directional aerial not only makes the tuner more sensitive but enables one transmitter to be singled out from interference from others, giving better audio quality. Also a distant stereo transmitter can be received rather than a local mono one if the aerial is aimed correctly. Get advice from a local dealer, however, as to what is best in your area; it varies enormously.

When improvements are required and the budget can stand them, the order to buy them in is as follows:

1. *Pick-up arm.* This will help preserve your record collection.
2. *Cartridge.* This will also help, but the chances are that the old pick-up arm would not stand a better-quality cartridge.
3. *Record deck.* This should improve performance figures. A more expensive deck is usually easier to handle, enabling records to be better looked after. (It is often easier to buy items 1 and 3 first and at the same time as it is tricky to change the arm alone. A new deck will often have its own arm fitted already.)
4. *Loudspeakers.* It is rare that the first speakers one buys are the best available, but given the percentage-of-cost rule of thumb outlined above it is probable that little improvement will be heard until this stage. Even if it were, it is better to avoid record damage first: fifty records probably cost about £150.
5. *Amplifier.* A cheap amplifier probably has restricted facilities and output power, but otherwise

Choosing a Hi-Fi System

its performance should be adequate until the rest of the equipment is improved. After using your hi-fi for a while, it will be easier to see which extra facilities are going to be useful to you and which are just more knobs to twiddle, so the experience gained up to this stage will help save money. Higher output powers do not necessarily mean that the hi-fi must be louder but do mean that short peaks in the music will not be distorted through inability to handle them.

6. *Other pieces of equipment.* Usually a tape recorder or radio tuner will have to be totally replaced if a better model is required. The assumption is that it is easier to save up for something expensive if you already have a substitute which works, even if not so well.

This list is compiled to enable expansion and improvement to be planned sensibly at the outset on the principle that it is wiser to spend less money at the outset and find out how far one's tastes and sensitivity develop, than to buy a very expensive set of equipment and never realise its full potential. If you buy a cheaper hi-fi system, then more records or tapes can be bought at the outset and this usually leads to greater pleasure for your money.

FREQUENCY RESPONSE

Frequency is defined in more detail in Chapter 10, but low notes are low frequencies and high ones are high frequencies. The frequency response of a piece of equipment is often shown as a graph with frequency as the horizontal axis, and the amount of output for a given size of input as the vertical axis. This latter is expressed in decibels (dB, defined in Chapter 10) which is a scale which takes into account the way the sensitivity of the human ear behaves. The point arbitrarily taken as a

Choosing a Hi-Fi System

reference level is 0 dB, and is usually defined at 1 kilohertz (approximately two octaves above middle C in the musical scale). Positive decibels mean the output is bigger, negative that the output is smaller. There is no absolute value for 0 dB: it has to be taken in context.

Human hearing, depending on the individual, extends from above 20 Hz in the bass to less than 20 kHz in the extreme treble (this is called its "bandwidth"), so most hi-fi equipment will have a response from 20 Hz to 20 kHz so that nothing that the listener can hear is lost. This range is, in musical terms, approximately $3\frac{1}{2}$ octaves below middle C to $6\frac{1}{2}$ octaves above.

The frequency response of a system is defined by the narrowest response it contains, so if an amplifier has a bandwidth from 20 Hz to 20 kHz, a pick-up cartridge from 30 Hz to 12 kHz, and a loudspeaker from 70 Hz to 17 kHz, the overall system will have a response from 70 Hz to 12 kHz. Do not be too discouraged if the system you choose has a bandwidth less than ideal; the difference is not terribly noticeable and to increase it costs rather a lot. There is a law of diminishing returns on the improvement in equipment for increase in cost.

As the ear is the final judge of a piece of hi-fi equipment, a listening test should always be made, if it is possible to do so, before buying anything. It does not matter how good the specification is; this is a guide as to the quality which enables some kind of choice to be made, for if it sounds wrong it is wrong.

DISTORTION

Distortion is a measurement of how well the quality of the input to a piece of equipment is maintained. Distortion should be low, of the order of 0.1 per cent. (This is discussed in more detail in Chapter 10.) This figure is just about the threshold above which it can be heard, and below which the ear will not detect it. Distortion can

Choosing a Hi-Fi System

give rise to listening fatigue, which is a sort of irritation one gets after listening to distorted music for an hour or so; one has just had enough. This can occur even though the distortion was not definitely audible in the short term.

STEREO

Stereo is a technique for making sounds appear to originate over a space between two loudspeakers. The signals fed to the loudspeakers are different from each other, though they bear a relationship to each other which defines where the sound appears to come from. A stereo sound system therefore consists of two electrical systems, or channels, sharing the same controls so that they are always matched, and contained in the same boxes as each other. The fact that the two systems are matched does matter or the sound picture gets changed for the worse. It is not simply a case of high notes coming from the left and low ones from the right, though this is the way most symphony orchestras are laid out; stereo can make any sound come from anywhere between the speakers that the record (or radio) producer intended. Records, radio and tapes have the signal coded on to them so that the hi-fi system can extract them and reproduce the two channels separately, despite the fact that there is only one record groove, radio station or tape being used as a source. How this is done is explained in the following chapters, and it is standard for each medium.

QUADROPHONY AND AMBIOPHONY

Quadrophony is a similar technique, but four channels are used to drive four loudspeakers. The coding of information on to records, radio, etc., is more complex, and so is the equipment used to decode it to drive the speakers. At the time of writing, there is no universal

Choosing a Hi-Fi System

standard as to how this coding is done, and which system will be finally adopted is anybody's guess. The end product of quadrophony is that the sound appears to come from all round the listener, re-creating a complete environment quite successfully if the record has been well made.

Ambiophony is a cheaper way of providing surround sound by starting with stereo and distributing the signals between four speakers in such a way that the main signal comes from speakers in front of the listener, and signals which give depth to the sound come from behind him. Used with certain records this can produce an effect nearly as convincing as quadrophony, but the results are very variable from record to record.

COMPATIBILITY

Compatibility has two meanings. When used of stereo signals, it means that if a record, for instance, is mono compatible it will sound all right played on a mono system – provided the correct cartridge is used, *i.e.* one that is stereo compatible (*see* Chapter 3). Not all records are mono compatible, though all can be played on a mono system, since some stereo effects used in pop music disappear completely in mono. Record manufacturers usually take care where possible to ensure mono compatibility. Similar definitions exist for quadrophony to stereo, etc.

The other use of the word "compatibility" is to describe whether one piece of equipment can be used with another. A hi-fi system will be only as good as its poorest component, so a perfect amplifier used with a distorting record player will reproduce the distortion beautifully, but it is still distortion. In improving a system this may mean that the benefits of one improvement can be heard only as the rest of the system catches up. This was taken into account in the ordering of the list earlier in this

Choosing a Hi-Fi System

chapter. This means that all the parts of a system should be of a similar quality.

It is important that the various parts of a system should be electrically compatible; that is that the output of one piece can drive the input to the next, and the input of the next can handle the output of the first.

Firstly record player cartridges have to have special amplifier inputs. A magnetic cartridge needs one type, a ceramic cartridge another. These are usually specified in the handbook of the amplifier. Secondly loudspeakers must be of the correct impedance for the power amplifier; this figure is expressed in ohms (Ω) and is not a figure of merit but an electrical property. If their impedance is too low there is a risk of damaging the amplifier; if it is too high the amplifier will not give its full power output. Conversely, the loudspeakers must be capable of handling all the power the amplifier can deliver or the amplifier may damage them. Power is measured in watts (W).

All other connections between pieces of equipment should obey the following rules:

1. The output level of one part should match the input sensitivity of the next. These are both measured in volts (V) or millivolts (mV), where 1000 mV equals 1 V. A discrepancy of about three to one in these figures is usually acceptable, but *not* in the case of loudspeakers above.
2. The input impedance of a piece of equipment must be as high or higher than the specified loading capability of the output driving it. These figures are usually given in kilohms (KΩ) or megohms (MΩ), where 1,000,000 Ω equals 1000 KΩ equals 1 MΩ.

If these rules are followed then there is no reason why a tape recorder should not be fed into the radio input of an amplifier; the name "radio" is given to it merely because

Choosing a Hi-Fi System

that is what it would normally be used for, so the title is convenient. Obviously the legends on switches are more useful if the inputs can be used as they were intended to be.

TEST RECORDS

A system can be tested using a test record, of which there are several on the market. Use them according to the instructions that come with them. Do not be disappointed if your system is not absolutely perfect according to the tests, but make corrections where you have the knowledge to do so and remember that nothing is perfect.

Summary

1. Check that the system you want will be practical in the rooms you will use.
2. Look for component parts that bear the right cost relationship to each other.
3. Consider ease of improvement.
4. Check frequency response (ideally 20 Hz–20 kHz).
5. Check distortion (ideally less than 0·1 per cent).
6. Are all component parts electrically compatible?
7. Remember to get the correct connecting cables.

2 | THE ELEMENTS OF A HI-FI SYSTEM

The various parts of a hi-fi system each have a definite function, and, in general, all are necessary. They drop into four groups – signal source, preamplifier, power amplifier and loudspeakers – though some or all of these may be constructed in the same box for convenience.

The signal source, be it record player, tape recorder or radio tuner, takes some form of coded signal and turns it into an electrical signal which can be amplified to drive loudspeakers. The output of the signal source is usually of low power and cannot be used to drive a loudspeaker directly. In the case of a record deck it also has a non-linear frequency response which must be corrected.

The preamplifier is the next stage of the equipment. It corrects any intentional non-linearities in the signal's frequency response. It is also at this stage that the tone controls operate, allowing the listener to adjust the tonal quality of the music to his own taste, and usually to filter out unwanted high- and low-frequency effects – scratches or rumbles on the record. At the end of the preamplifier, the signal is still rather weak, but would normally be adequate to feed to a tape recorder for recording purposes.

The power amplifier follows the preamplifier, and has one function only: it converts the weak signal into one

The Elements of a Hi-Fi System

that is powerful enough to drive a loudspeaker. It is important that this amplifier is of good quality, as it must faithfully copy its input signal at the higher output level – it is usually at this stage that distortion (which can be simply avoided) is introduced to the signal. Loudspeakers are rather inefficient devices, so the amplifier must be quite powerful to produce high volumes. If the output is accidentally short-circuited it should protect itself, or it will absorb the output power in the form of heat, probably destroying some of its component parts.

The power amplifier drives the loudspeakers. These convert electrical energy into sound waves in the air. Considered at its simplest, a loudspeaker is a diaphragm which is vibrated by an electro-magnetic system; and this is exactly what early loudspeakers were. However, there are many limitations to the simple system, and modern loudspeakers have tended to become more and more complex to overcome these.

In describing the four sections of a hi-fi system, it was simplest to consider the signal source as a block in relation to the other three parts. However, there are distinct component sections within the sources themselves.

The record player consists of three basic parts: the turntable, the pick-up arm and the cartridge. Again, these may be supplied together as a single unit. The turntable is a disc 10 or 12 inches in diameter with a constant speed motor, and a gearing system to rotate it at $33\frac{1}{3}$ r.p.m. or 45 r.p.m. The pick-up arm is a balanced beam usually approximately 10 inches long, pivoted near one end, with the other end free to move in an arc across the record roughly along a radius. On the free end of the pick-up arm is mounted the cartridge, which consists of a stylus mounted so that it follows the vibrations of the groove in the record as the record slides under it, and a device to turn these vibrations into an electrical signal. This signal passes along very fine wires

The Elements of a Hi-Fi System

in the pick-up arm, and is then connected to the preamplifier input.

The radio tuner is normally supplied as a single unit, though it will probably need an aerial. The tuner selects the radio signal according to its tuning and decodes the audio signal from it. The output is usually greater than that from a record deck, and so will be fed to a less sensitive input on the preamplifier.

The tape recorder also comes as a single unit. It consists of a tape transport mechanism, the purpose of which is to pass the tape at a selected constant speed past the tape heads while it is playing, or to move the tape swiftly in order to select part of it to play, and also the recording and replay circuitry. The output is similar in amplitude to that of a tuner.

There is a choice with tape recorders between "open spool", cassette or cartridge systems, but the merits of each system are discussed in a later chapter.

Finally, if personal listening is required, headphones may be used in place of loudspeakers. These are basically very small loudspeakers mounted on a band so that they fit over the head with a speaker next to each ear. As they are so close to the ear, and tend to be more efficient than loudspeakers, they require much less power. Typically one-hundredth of a watt would give an approximately equivalent volume to a loudspeaker in a normal room.

3 | THE RECORD AND THE RECORD PLAYER

The basic essentials of a record player were described in Chapter 2, but, although that description would enable someone to recognise the three component parts of a deck, he would have little information on how to go about choosing one to match his pocket, or his ears.

THE RECORD

Before explaining why a record player is made as it is, and the considerations to be made in choosing it, it is probably wise to describe what the record itself is. The groove that runs from the edge of the record towards the centre has minute fluctuations in it from side to side, and it is in the shape and size of these that the sound waveform is stored. At the edge of the record, the groove travels under the stylus at approximately 50 cm per second, while at the end of the side it has reduced to just over 20 cm per second. If near the edge there are twenty fluctuations of the groove from one side to the other and back in every centimetre, then the stylus which is following the groove will vibrate at a rate of 1000 cycles per second. This produces a tone approximately two octaves above middle C. For accurate reproduction of the original sound both in pitch and at the correct pace, it becomes obvious that the record must go at an absolutely constant speed, and this speed must be the

The Record and the Record Player

same as that at which the record was originally recorded.

So far we have a record with one channel of sound on it in the form of a groove moving from side to side. Stereo sound requires two channels running together in time but separated as far as their content is concerned. It would be possible to record the second channel as an up-and-down movement of the groove, but this is not practical. Old mono record players have styli that will not move vertically very fast – 10 cycles a second is too fast.

This means that not only would a mono player just reproduce the channel recorded horizontally but also the other channel would be destroyed by the heavy action of the stylus. The ideal system should work as well with mono as it does with stereo equipment.

The solution is to record the channels one on each side of the groove, and with their fluctuations at right-angles to each other (*see* Fig. 1). It can be seen from the diagram that channel B can move back and forth along the direction of its arrow (see dotted lines) without channel A being affected and vice versa. If both walls of the groove move equally in the direction of the white head on the arrow, the effect is to move the groove sideways – this is the same as a mono signal.

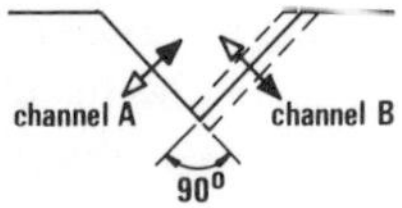

FIG. 1. Schematic of record groove

If a stereo record recorded in this way is played on a mono cartridge (that is one that responds only to horizontal vibrations) it can be seen that the stylus will pick up channel A and channel B and effectively add them together. Figure 2 shows only the channel B part of this process.

The problem of vertical movement is still not overcome,

The Record and the Record Player

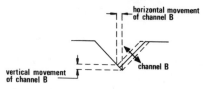

FIG. 2. Effective vertical and horizontal movement of record groove

and for this reason a "stereo compatible" mono cartridge should be used to reproduce stereo records in mono. These cartridges allow the stylus to move both horizontally and vertically, but only the horizontal signal is produced electrically. The other solution is to use a stereo cartridge, and connect it up so that channel A and channel B are added to each other to produce a mono signal. If it is planned to convert a mono system to a stereo one at a later stage, this solution is probably wiser as it saves buying a different cartridge when the change is made.

In order to play a stereo record in stereo we now need only to have a cartridge with two outputs; one of which responds to a vibration at 45° to the horizontal in one direction but not the other, and the other at 45° in the other direction but not the first.

THE CARTRIDGE

There are two methods of producing an electrical signal from the vibrations of the stylus in general use at the moment. The first is a crystal, or ceramic cartridge. This works on the principle that piezo crystals produce a voltage if they are bent. Two of these crystals shaped as small beams are mounted as shown in Fig. 3. It can be seen from this illustration that one crystal is bent by the vibrations of one channel and the other crystal by the vibrations of the second channel.

If the crystal on the right is bent, however, the one on the left is also being twisted and so generates a small

The Record and the Record Player

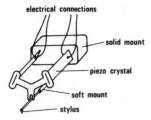

FIG. 3. Schematic of inside of ceramic cartridge

voltage. The isolation of the signal between channels is therefore not very good. Elegant solutions have been found to this and similar problems.

In its favour, the ceramic cartridge is cheap, it is rugged and it produces a large output voltage. Against it, however, is the fact that it has to be connected to an amplifier with a very high input impedance – easy with valves, but not so with transistors for engineering reasons. The ceramic cartridge is also prone to distortion, is rather noisy (*see* Glossary) and is difficult to manufacture with high compliance (*see* below). Although solutions exist to these problems they take the cost of a ceramic cartridge higher than that of a magnetic one of similar performance.

The magnetic cartridge uses the fact that a voltage is generated in a piece of wire if it is moved across a magnetic field, or that if the magnetic field through a coil is altered a voltage is generated across it. One such system is shown in Fig. 4, but there are many types possible.

Again, one of the coils responds to its channel alone, and the other coil responds only to the second channel. However, it is easier to separate the electrical effects between the channels. In the magnetic cartridge's favour are not only better isolation but also lower distortion, higher compliance and lower effective mass of the stylus (*see* below). It also requires a lower input impedance on the amplifier.

The Record and the Record Player

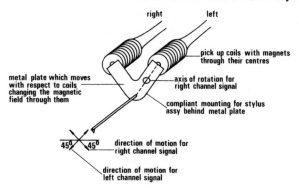

FIG. 4. Schematic of inside of magnetic cartridge

Against it are higher cost of manufacture, higher total mass of the cartridge and low output voltage.

Because of the higher compliance, the magnetic cartridge can be used at a lower playing weight than the ceramic one, and so is gentler to records; but this is also subject to the quality of the pick-up arm which is discussed later. However, it is generally preferable in a good hi-fi system to use a magnetic cartridge.

Most of the cost of a magnetic cartridge is in the stylus, and may be up to 80 per cent despite its small size. This means that some manufacturers do not make the stylus removable, and when it becomes worn out the whole cartridge is replaced.

THE STYLUS

The stylus is on a cartridge as the needle was on a gramophone. It consists of a mounting system of some type, and a thin bar about a centimetre long with a diamond or sapphire chip on the end of it. Better-quality styli have diamond tips, as these last longer than sapphires. The tip will therefore be referred to as the diamond from now on.

The job of the stylus is to pick up the fluctuations in the

walls of the record groove, and transmit them as vibrations to the electrical system in the cartridge. It therefore must move very easily (high compliance) but the actual bar should not bend or some of the signal will be lost in transmission. Neither should the stylus resonate, or vibrate more easily at one frequency than at others, as this would exaggerate that frequency and produce an uneven frequency response.

Although different qualities of styli look much the same, the higher-priced ones are more delicately made, and overcome difficulties of compliance and resonance. As we would expect, high-compliance cartridges require a lower tracking weight – that is the amount of force holding the diamond down against the record. Firstly the diamond follows the groove more easily, and so does not need to be pushed into it so hard; secondly, if too much force were applied, the tip of the stylus would be pushed up into the cartridge itself and the casing would rub against the record.

No point is infinitely sharp, and this is true of the diamond on the stylus, which is intentionally rounded (*see* Fig. 5).

FIG. 5. Tip of stylus

There is a choice of two shapes for the diamond, known as "spherical" and "elliptical". On a spherical diamond, the tip is a minute hemisphere with a typical radius of 0.0007 in. The elliptical diamond is wider than it is front to back (0.0007 in. wide, 0.0002 in. front to back typically) and it is more expensive. The reason for using it is as follows (*see* Fig. 6).

The record is first cut using a stylus shaped like a chisel

The Record and the Record Player

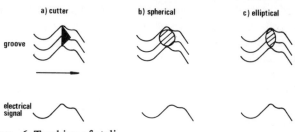

FIG. 6. Tracking of styli

on a cutting lathe. Figure 6(a) shows the signal to be recorded under a sketch of the cutter in the groove – the cutter being triangular, and the record groove moving in the direction of the arrow. Ideally the stylus used for playback would be identical to that used on the record lathe, but this would also act as a chisel, and quickly wreck records.

Figure 6(b) shows a spherical stylus used to play back the same piece of the record, and it is obvious that it simply won't fit into the little kink at the top. This corresponds to a high-frequency signal, and it will be lost on playback.

Figure 6(c) shows an elliptical stylus playing the same section, and it can be seen that it will produce an output which is very much the same electrical signal as was originally recorded.

Referring back to the earlier comments about chisels, it is obvious that any stylus pressing on the record much too hard will act as a cutter, and so the playing weight should be kept as low as possible. Most cartridges are supplied with a manufacturer's note stating the range over which they operate – do *not* exceed this even if the record is throwing the stylus out of the groove as it is almost certain to damage the cartridge, and will certainly damage the record; look for the cause elsewhere, probably in the pick-up arm.

If the cartridge is used underweight there will be distortion on loud passages of music as there is not enough force holding the stylus in the groove and it will jump very slightly clear of the groove walls, giving incorrect signals.

Within the operating range of playing weights that the manufacturer recommends, there is usually a trade-off of frequency response against channel separation or crosstalk. At the heavy end of the scale the frequency response will be better, whereas at the light end the separation will be better. It is probably a good idea to run a high-compliance cartridge near the heaviest playing weight for which it is designed – it still will not damage the record. However, as the quality of the cartridge is decreased, one should tend to use it nearer the light end of the scale.

All this assumes that the pick-up arm is good enough to carry the cartridge. A cheap deck and arm will not be improved by fitting a professional cartridge costing twice as much as the deck.

THE PICK-UP ARM

Basically the pick-up arm is, as described in Chapter 2, a beam pivoted near one end, and just off balance so that a known force presses the cartridge mounted on the other end against the record. The cartridge end can move up and down, and also in an arc across the record from the edge towards the centre.

Records are made with the cutting stylus moving on a radius across the record, so the swinging track of the cartridge is a compromise between convenience and the ideal path which would copy the cutting lathe (*see* Fig. 7).

Parallel tracking arms are designed to overcome the fact that only at one point on the record is the centre line of the cartridge actually parallel to the groove being

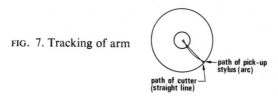

FIG. 7. Tracking of arm

played as was the cutter during the making of the record – near the edge it will point outwards and near the centre inwards (*see* Fig. 8, where the tracks are drawn as straight lines and the angles exaggerated for clarity).

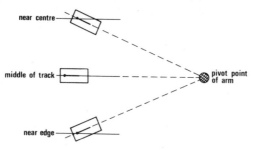

FIG. 8. Non-parallel tracking

There are two basic types of parallel tracking arm. One type keeps the cartridge's angle correct with respect to the groove, but still moves in an arc. This is done by a system of levers. The other moves the whole arm and its pivot across the record with a system of motors so that the cartridge moves in a straight line along a radius. This emulates the cutting lathe. Both of these systems are expensive: the intending purchaser will have to decide whether the benefits conferred are worth the extra cost. Remember that a higher degree of automation is available with this type of player. Distortion occurs at the end of the side of a record, but this is as much to do with the lower groove speed mentioned earlier as with tracking errors.

The Record and the Record Player

At this point it is probably wise to mention "moment of inertia". This is the result of multiplying a weight or mass by the square of its distance from a pivot, and it is a measure of how easily something may be accelerated. Hence the moment of inertia of a pick-up arm is the measure of how easily it will follow changes in the rate at which the groove on a record spirals towards the centre. In between individual tracks, for instance, the arm has to swing quite quickly from the end of one to the beginning of the next. Also, if a record is warped, the arm must swing up and down fairly rapidly as the record rotates. There is no driving force on the pick-up arm other than the stylus following the groove and pulling the arm with it. If the arm has a high moment of inertia this will be difficult to do, and a delicate stylus might become damaged by the forces involved. Certainly it is possible to damage the record at points where the arm has to change the speed at which it is moving. Hence a pick-up arm should have as low a moment of inertia as possible. The heavy balancing weights are very close to the pivot, so they have little effect. However, the headshell (the section at the end which actually carries the cartridge) and the cartridge are a significant distance from the pivot and quite small changes in their weight will alter the moment of inertia considerably. For this reason it is wiser to use an arm with a fixed headshell as the clamping arrangement and socket of a removable headshell greatly increase the moment of inertia. However, this is a point in choosing a pick-up arm where technical perfection may be traded off against convenience, for if it is required to move the deck from place to place the cartridge should be removed each time a move is made. This is obviously easier if it is simply a case of removing the headshell.

As with styli, the pick-up arm should not resonate. The cheap solution to prevent this is to build a very solid arm, but this raises the moment of inertia and so a less compliant cartridge must be used with a higher playing weight, giving risk of damage to records. This damage may not be heard on the cheap record player, but if the records are played on a better one it will be easily discernible and often quite unpleasant. It seems extravagant to have to change what might be quite a large and expensive record collection simply because a better record player has been bought and its predecessor was damaging records.

Finally in this section, the pick-up arm should have some way of applying a gentle force that tends to move it outwards on the record – this is known as "bias". The reason for this is that, owing to friction, the record tends to try to pull the stylus round with it, which exerts a small force on the arm. This force does not pull directly on the pivot point, but slightly to the right of it (*see* Fig. 9).

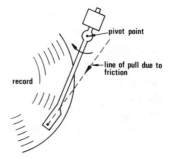

FIG. 9. Bias force requiring compensation

This tends to turn the arm in the direction of the dotted arrow, resulting in a higher playing weight on the inside edge of the groove than that on the side nearer the edge of the record. The concept of stereo requires that the two channels should be played under as near as possible identical conditions, so an opposite force must be applied to counter this tendency. If the bias is badly adjusted the buzz characteristic of mistracking may be clearly heard from one of the channels.

The Record and the Record Player

Some manufacturers build in bias compensation which is set as a reasonable compromise for the requirements of record speed, modulation, stylus contact pressure and tip shape. It is arranged to be non-adjustable by the user, so simplifying the setting-up of the arm.

THE TURNTABLE

The only function that a turntable performs is to rotate the record at a constant known speed under the stylus, but it must do this job well, as slight imperfections in performance can be very audible. It is probably best to describe the imperfections first to make the later explanations simpler.

Firstly it is important that the speed of the turntable is the same as that at which the record was cut. There are two international standards in common use today, $33\frac{1}{3}$ and 45 r.p.m.; 78 r.p.m. used to be used for records before modern techniques enabled the speed to be reduced and the groove made finer so that a symphony, for instance, could be recorded on one record instead of six or more. If the speed is wrong the pitch of records is changed, which can be disturbing to someone with perfect pitch. However, the speed can be checked with a stroboscopic disc. This device uses the fact that artificial lighting flickers one hundred times a second. In one-hundredth of a second, a turntable rotating at $33\frac{1}{3}$ r.p.m. will have moved a one-hundred-and-eightieth of a revolution. If a disc with 180 equally spaced marks round its edge is put on the turntable, then each time the light reaches a peak in brightness each mark should have moved into the position occupied by its neighbour at the last peak. The human eye will detect this, and a stationary pattern will be seen. If the turntable is going too fast, the pattern will seem to move clockwise; if too slowly it will move anticlockwise. The same can be done at 45 r.p.m., except that there are now roughly 133 marks, but because the arithmetic is not quite right the turntable will be going one quarter of a per cent too fast.

There are two related faults called "wow" and "flutter" which are caused by slight variation in speed. Wow is usually due to slight imperfections in the drive system and it happens slowly, usually rising and falling once or twice a revolution of the turntable. It is usually really noticeable only on sustained bass notes, particularly in piano or organ music. Wow can also be caused by a record having its hole just off centre, in which case the turntable is not at fault.

Flutter is a very rapid speed fluctuation, and it shows up as a warble, again particularly on sustained notes, but not necessarily in the bass. This is caused by the fact that no motor drives absolutely smoothly, and that the speed fluctuations are getting through the drive system to the turntable. As a way of counteracting this, the turntable itself is designed as a flywheel, and the motor is connected to it either through a soft idler wheel or with a rubber drive belt, so that the small, fast motor speed fluctuations are not transmitted. It is the moment of inertia of the turntable that matters here to determine how good a flywheel it is, not necessarily its weight, though this is usually a fairly good guide. Again there is a trade-off of design points, for if the motor is less powerful it requires a less heavy flywheel to smooth out any flutter there may be. However, a minimum power of about 3 watts is necessary to start the turntable and overcome friction without straining the motor. Also, the softer the link between the motor and the turntable the better, within reason, and the drive-belt system is better controlled and less prone to ageing than that using an idler wheel.

Lastly there is the possibility of "rumble" in a turntable; this is actual mechanical noise generated usually in the bearing at the centre. For this reason in particular,

The Record and the Record Player

ballbearings (or needle bearings) are not used since the balls, in rolling round, tend to vibrate the turntable, and the stylus picks up the noise they make. Rumble may in fact be recorded on to records, as it is present in all the mechanical systems used to make them, so again the turntable may not be at fault.

SETTING UP

Normally a manufacturer will supply information on how to set up a record deck, However, generalised procedures will be described here. These in no way supersede the manufacturer's instructions:

1. Check that the mains supply voltage setting of the deck and its mains frequency requirements are correct for your area. In Britain these will be 200–240 V at 50 Hz.
2. Mount the deck on a solid plinth (though this may be supplied with it) and place this on a solid base which will not be shaken by footsteps on the floor near it, and preferably not by people touching it either. Particularly at low playing weights the stylus can easily be knocked out of the groove by a shock transmitted through a table to the deck. This usually leaves a click on the record, if not a scratch.
3. Make sure that the deck is level, preferably with a spirit level.
4. Reduce the bias compensation to zero: see manufacturer's handbook.
5. Fit cartridge and stylus to the arm if this is not already done, and, if there is a detachable stylus cover, remove it. Do not tighten the cartridge securing screws fully, but just enough to grip the cartridge allowing it to move slightly. If the international colour code is used for the connecting wires to the cartridge, it is as follows:

White: Left Signal
Blue: Left Ground

Red: Right Signal
Green: Right Ground

6. If the manufacturer has supplied a gauge to position the stylus correctly, then use it according to the instructions. If not, various gauges are available from shops. The purpose of doing this is to get the stylus into the best compromise position to copy the cutting stylus position when a record was made.

7. Place the arm over the turntable in its raised position, and approximately halfway between the edge and the centre of the turntable. Without altering the distance of the stylus from the pivot of the arm, angle the cartridge so that its centre line is tangential to the line of the groove at this point. Tighten the cartridge-securing screws.

8. Balance the arm according to the manufacturer's instructions, and if possible check any tendency in the arm to swing sideways. It may be necessary to raise one edge of the deck to do this, but, if this is done, make sure that the deck is stable. Apply the correct amount of playing weight for the cartridge you are using.

9. If adjustment is available, apply bias compensation to the arm according to the manufacturer's instructions and play a record. If after this there is a distinct buzz on one channel only, first check that the system is clean, then change the bias a little at a time (but not while the record is playing), increasing it if the buzz is on the right channel, decreasing it if it is on the left. The buzz implies that the stylus is resting too lightly on that channel; the outer edge of the groove is the right channel and the inner the left, and the bias compensation pulls the pick-up arm outwards.

If the buzz occurs on both channels together it is possible that the playing weight is too low – but do *not* increase it above the maximum specified for the cartridge.

The Record and the Record Player

Summary

1. Ceramic cartridges are cheap; magnetic cartridges are higher quality. Check the amplifier has an input suitable for the one chosen.
2. Elliptical styli give higher fidelity than spherical ones but tend to be more delicate. If the record player is to be used by all-comers it is probably better to get a more rugged stylus. There is no reason not to have two cartridges on different headshells and keep the better one for "private" use on valuable records.
3. The pick-up arm should be long with low inertia.
4. There should be bias compensation, preferably adjustable.
5. A heavy turntable is usually a good idea, though modern techniques are making this less critical.
6. Set up the record player correctly, or distortion may result.

4 | THE RADIO TUNER

As with making records, radio transmission is a way of encoding an audio signal, only the purpose is to send the signal a long way rather than to remember it. The universe contains minute electrical and magnetic forces which bear a fixed relationship to each other and, seen at one position, do not naturally alter much. However, if an electrical disturbance occurs, it spreads outwards from its source as oscillations in the balance of these forces, in much the same way as ripples spread from a stone dropped in a pond. A radio transmitter's aerial is one way of causing just such a disturbance in a controlled manner (lightning is another, but rather less controlled!).

Like an organ pipe, a given length of aerial tends to oscillate best at a given frequency, and the transmitter drives it with a signal at that frequency. The effect is that electrons rush up and down the aerial, dragging the electrical field near the aerial with them and altering the balance of the electro-magnetic field, that is the pattern of the electrical and magnetic forces. If, at some distance from the transmitter, a piece of wire is hung up, the electrons in that wire will move in sympathy with oscillations in the field near it.

There is a specific length of wire which is best for picking up a given radio frequency (RF), but although it tends to pick up that frequency it will also pick up others.

The Radio Tuner

The first job of a radio receiver is to filter out unwanted signals, and select just the one required. Most modern receivers do this by basically the same method.

The incoming RF signal is roughly filtered as soon as it enters the tuner and it is then heterodyned. Heterodyning causes the whole range of incoming signals to be changed down in frequency, but it does not alter the information encoded on those signals. The output from the heterodyne circuit is fed to a very selective filter which will respond only to one signal frequency. This is known as the "intermediate frequency", or IF. By altering the amount by which the incoming signals are shifted, the one which produces the correct IF may be changed, thus tuning in different stations. The reason for this complex procedure is that it is very difficult and therefore expensive, from an engineering point of view, to produce a filter whose frequency can be changed, but which is selective enough to pick out just one radio station.

The rest of the radio tuner's circuitry now decodes the selected signal to produce the audio frequency signal it contains. This is known as "detection", and in the case of a portable receiver the signal is then amplified to drive a loudspeaker. There are two ways in general use of encoding the audio on to the radio frequency: amplitude modulation (AM) and frequency modulation (FM). AM is used principally for short-, medium- and longwave transmission (SW, MW and LW); FM is used for "very high frequency" or VHF (UHF, used for 625-line television, stands for "ultra high frequency").

Amplitude modulation means that the amplitude of the transmitted radio signal is altered (modulated) by the audio waveform (*see* Fig. 10).

To extract the audio frequency signal from the IF signal generated in the tuner (which mimics the RF signal in amplitude), the detector circuit ignores the actual IF,

The Radio Tuner

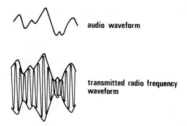

FIG. 10. Amplitude modulation

but detects how big it is at the time, so reproducing the audio signal originally transmitted.

Frequency modulation uses an RF wave of constant amplitude, but the frequency is altered according to the audio waveform. The peaks on the audio wave increase the frequency, and the troughs decrease it from its basic frequency (*see* Fig. 11, in which the effect is exaggerated). In fact the changes in frequency are so slight that they are not lost in the filter of the IF stage.

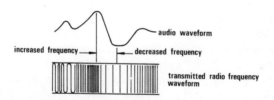

FIG. 11. Frequency modulation

The detector on an FM tuner is called a "discriminator" as it discriminates between frequencies.

Because of the slight frequency changes, the tuner of an FM set must not change its tuning with time or temperature, or the selected station will be lost and the tuner must be re-tuned to catch it again. Fortunately, because of the way it works, if the tuning drifts the average voltage at the output of the discriminator will

The Radio Tuner

drift as well. This produces an electrical signal which can be used to return the tuning to its original setting. The operation of this type of circuit is called "automatic frequency control" or AFC. It should be possible to switch the AFC on and off since it will try to prevent the tuning knob from having any effect, thus making it difficult to tune into another station. In fact the AFC has a much more limited range than the tuning knob, so it will always be possible to change stations. However, it may be a rather sudden effect as the AFC lets go of one station and grabs on to the next – it may even jump right across the next station, if the signal from it is weak, and catch the one beyond it on the tuning scale.

With both AM and FM, the actual strength of the received radio signal will vary with the distance from the transmitter and with atmospheric conditions. Because of this, a tuner will normally contain an automatic gain control (AGC) circuit which measures the average amplitude of the IF waveform, and adjusts the amount by which the signal is amplified so that the average output of the IF circuit is reasonably constant. This is important as otherwise the quality of the audio will depreciate the further the receiver is from the transmitter.

If an FM tuner is not tuned to a station, it produces a crackling hiss which is an amplified form of the slight RF noise which always exists in the background but is normally much quieter than a radio transmission. When this happens, the AGC circuit is operating at or near its limit, and a circuit can be included to detect this. The output from this circuit is used to switch off the audio output so that there appears to be silence between stations. This type of circuit is called "inter-station muting", or, more descriptively, a "squelch circuit".

A block diagram of an FM radio incorporating all of these features is shown in Fig. 12.

The Radio Tuner

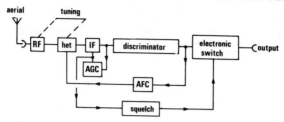

FIG. 12. Block diagram of FM receiver

STEREO RADIO

It is possible to transmit both channels of a stereo signal on one frequency of radio transmission, but it requires a rather wide bandwidth, that is the difference between the highest and the lowest frequency used in the transmission. Even with AM, although the radio frequency is constant, the addition of the coded signal to it involves use of frequencies on either side of it. If the full audio bandwidth of 20 Hz to 20 kHz were transmitted on an RF signal of 100 kHz, the radio signal would occupy the band from 80 kHz to 120 kHz and make that band useless for other transmissions. If, however, it were transmitted on an RF signal of 1000 kHz (1 MHz) it would occupy from 980 kHz to 1020 kHz. It can be seen that the higher the RF or carrier frequency, the lower is the percentage of the carrier used for a fixed audio bandwidth. Similar rules hold for FM.

The medium wave (MW) band, which is principally used for domestic transmissions, is already cluttered, even though the audio bandwidth transmitted is limited to $4\frac{1}{2}$ kHz instead of 20 kHz. To open a station transmitting stereo, and therefore requiring more bandwidth, would be impossible. To use two stations, one for each channel, not only would be inconvenient for the user who would have to use two radios to pick them up, but would also require some other service to be taken off the air. For these reasons among others, the VHF band from

The Radio Tuner

97 to 108 MHz is used, where the necessary wide bandwidth of 38 kHz is a negligible proportion of the carrier frequency.

The stereo signal consists of three sections which are chosen so that a mono tuner will still be able to receive the whole programme even if a stereo programme is being transmitted. The first part consists of the sum of the left channel plus the right channel (L + R), transmitted normally; the second is the difference between the left and right channels (L − R), and this is encoded so that it appears in a band above the frequency band occupied by the L + R signal. The third part is a pilot tone at 19 kHz which is used by the stereo decoder to recover the L − R signal. It is also used to activate a circuit which illuminates an indicator light if a stereo programme is being received, and sometimes automatically switches the tuner to stereo operation if this is so.

If the complete stereo signal is picked up on a mono tuner, it responds only to the L + R signal, as it does not have the equipment to utilise the rest; this signal is the mono version of the stereo being transmitted. Not only is the rest of the stereo signal out of the human hearing range as it is at too high a frequency, but also a tuner has built into it a circuit (called "de-emphasis") which filters high frequencies. A radio signal is exaggerated in amplitude at the treble end on transmission. The reason is that tuners are prone to noise at those frequencies and the signal strength must be considerably greater than the noise. When the tuner corrects the exaggeration it also diminishes the high-frequency noise, improving the quality of the audio signal (*see* Fig. 13).

A stereo tuner, however, produces within it the two signals, L + R and L − R. By adding them to each other it produces 2L, and by subtracting them it produces 2R. These are the left and right channels of the stereo programme.

The Radio Tuner

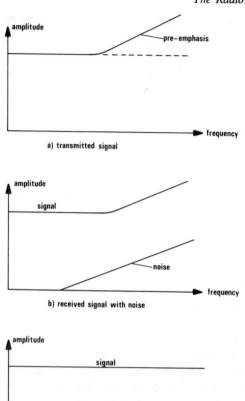

FIG. 13. Effect of pre-emphasis and de-emphasis

If the signal being received is rather weak, the stereo decoder will pick up a lot of noise of a similar signal strength, which makes it difficult to sort out the stereo signals. (There is no de-emphasis to reduce the high-frequency noise and make its job easier; this comes after the

The Radio Tuner

decoder.) As a result, stereo programmes from weak sources may be accompanied by an unbearable crackling hiss, and a stereo tuner should have a switch to enable the stereo circuits to be switched off as the mono signal will be of better quality.

AM OR FM?

The true high-fidelity transmissions in the UK are broadcast on the VHF FM band. Very nearly all the frequency range audible to the human ear is transmitted – only the top $\frac{1}{3}$ octave is missing, and many people cannot hear a tone that high anyway. It is only on the VHF stations that stereo is transmitted, though for a portable receiver that will never be connected to a hi-fi set this won't matter.

The actual extraction of the audio signal from the RF signal in an AM receiver tends to produce distortion which is not so prevalent in an FM tuner. However, AM receivers are cheaper – the FM circuitry has to operate at much higher frequencies, and is normally much more sophisticated. In the United Kingdom, although there is a certain amount of duplication between the AM services and the FM ones, certain programmes are missing from either band, and the choice of receiver must depend on which programmes you wish to receive.

In the choice of a portable set, a reasonable rule is to look for a radio which is in a solidly made box (some plastic ones are more solidly built than wooden ones, surprisingly). The box should also be as large as possible as the quality of the bass depends on this. Obviously, though, the choice of size of box must also depend on convenience.

AERIALS

The aerial on a radio tuner is its source of signal, and so

The Radio Tuner

should be good enough for the job. An expensive tuner is not worth the money if it is not receiving enough signal on its input to perform properly. Be prepared to spend extra money on an aerial, particularly in an area where the signal is weak.

As stated at the beginning of the chapter, a piece of wire might be used as an aerial, but it is not very efficient, and requires a very sensitive receiver when picking up weak signals. Near a transmitter, of course, it would probably be perfectly adequate.

AM receivers nowadays usually do not require an external aerial as they have one built in. This is in the form of a coil of wire round a piece of ferrite. The ferrite picks up the magnetic part of the electromagnetic waves, and the coil turns this into an electrical signal. For short-wave (SW) transmissions an external aerial may, however, be needed as these tend to be at a lower signal strength.

VHF signals tend to be weaker and so FM receivers will probably need some form of external aerial as the stronger the signal fed to the tuner the better the quality of the audio is likely to be. However, if it is too strong, the input circuits are likely to be saturated and the quality will go right down. Also, the inside of the tuner is probably screened to avoid spurious pick-up of unwanted signals, so the aerial socket may be its only contact with the outside world.

A piece of wire may work – after all the telescopic aerial of a portable set is just an elaborate form of this. The next stage in complexity is called a "dipole" (it has two "poles"), which should be half of a wavelength in length from tip to tip (*see* Fig. 14).

FIG. 14. A simple dipole

The Radio Tuner

For FM transmissions in the United Kingdom, this should be approximately 5 feet. If the aerial is made from 300 Ω feeder cable (bought as such from a dealer), the last $2\frac{1}{2}$ feet may be split in two, and the two wires bent to go in opposite directions as in Fig. 14. The path of the rest of the feeder to the tuner does not then matter, but it should be fed into the 300 Ω aerial input socket. It is a happy accident that the matching of the electromagnetic signal to the aerial, and the aerial to the tuner, will be correct if this is done.

Other aerials exist which are more and more efficient as the complexity increases. They also tend to become more directional, which may be useful if it is required to pick up a distant stereo station rather than a nearby mono one when the two frequencies are close together.

Alternatively a highly directional aerial may be used to pick up a distant set of transmitters (maybe 50 miles) with more programmes available rather than a local one with few programmes.

These aerials are normally given a gain figure in dB's, the higher the better (and the more expensive, often). This figure expresses how much better they are than a dipole operating in the same place.

Whichever aerial is used, it should be mounted as far away from obstructions as possible (trees, houses, gasholders, etc.). This usually means as high as possible. Also, even with a simple piece of wire and certainly with other types, it does matter in which direction it is pointed. Probably the best way to find the best direction to point the aerial is by trial and error, checking for the best signal quality (or for a maximum deflection on the tuning meter if this is fitted to the tuner).

The last words on aerials should be that they are not simple, and it is frequently less costly in the long run to get expert advice as to the best one to have for the area in which it is to be used.

The Radio Tuner

Summary

1. AM gives wide choice of mono stations at lower fidelity; FM gives hi-fi (and, increasingly, stereo) over a limited number of stations in the United Kingdom.
2. Additional controls on a tuner give better capabilities in general.
3. Check the sensitivity of the tuner is suitable for your area. A nearby transmitter requires less sensitivity.
4. Check if stereo transmission is or will be available in your area before buying a stereo tuner.
5. Make sure some of the budget available is allocated to buying a sensible aerial.

5 | THE AMPLIFIER

The purpose of the amplifier in a hi-fi system is to take the weak electrical signals from the various sound sources, for example a record player pick-up or a radio tuner, and increase them to powerful signals to drive loudspeakers. This process is called "amplification". Nowadays the techniques for controlling electrical signals are accurate and sophisticated, and this is why electrical amplification is used in preference to any other method. The horn of a mechanical gramophone is effectively a mechanical way of doing the same job, but it is inaccurate and tends to have a rather distinctive and unreal sound.

The amplifier also changes the content of the input signal in certain cases, where this is necessary, and can be used to change the tonal quality of the programme material to suit the listener's preferences or to remove undesirable effects.

These two functions of the amplifier are performed separately within the unit. Content and tonal quality changes are made first in the preamplifier section of the unit. The changed input signal is then fed to the power amplifier section where the power of the signal is increased but the content is not changed. Sometimes the preamplifier and power amplifier can be bought as separate units.

The Amplifier

RIAA EQUALISATION

The various sound sources in a hi-fi system tend to generate general electrical noise, particularly at high frequencies. The audio signals are usually increased at these frequencies so that they are stronger than the noise generated. If this exaggeration of high frequencies in the signal were not corrected, the programme material would sound brittle and tinny.

In the case of radio, the tuner itself will correct this, but for record players the preamplifier does it by applying what is called "RIAA equalisation". The equalisation circuits used reduce the high-frequency components of the audio signals which were exaggerated at the recording stage. In so doing, the high-frequency noise is also reduced proportionally and, since the signal was boosted so that it was stronger than the noise, the noise is generally acceptable after equalisation.

Equalisation is also applied to the low frequencies of the signal from a record player. The amount of side-to-side movement of the stylus of a record player that is necessary to produce a note of a given loudness is inversely proportional to the frequency of the note. The lower the frequency of the note, the greater must be the swing and the greater must therefore be the separation between grooves on a record so that they do not interfere with their neighbours. To enable records to have closely spaced grooves, and hence longer playing time per side, low frequencies are reduced in amplitude at the recording stage and have to be increased again by the equalisation circuits in the preamplifier. This makes record players very susceptible to pick-up of mains hum since this is at a low frequency and will therefore be increased by the equalisation circuits along with the low-frequency sound signals. The equalisation required for microgroove records, that is RIAA equalisation, is shown approximately in Fig. 15.

The Amplifier

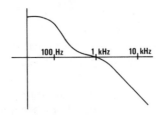

FIG. 15. RIAA equalisation

INPUT SELECTOR SWITCH

Most signal inputs to the preamplifier other than inputs from record players, usually labelled "Pick-up" or "Gram", will tend to be flat, that is they will have constant gain independent of frequency over the audio range. The input selector switch of the preamplifier connects the appropriate circuits to the input socket selected so that, for example, the signal from a magnetic cartridge of a record player, which is very weak, will be fed to a more sensitive input circuit with correct RIAA equalisation; whereas the signal from a radio or tape output, which is stronger, will be fed to a less sensitive input with a flat response.

TONE CONTROLS AND FILTERS

A flat response having been achieved, it may be required to change the tonal quality of the programme material. Raising the treble, or high-frequency, response will make it sound brighter; raising the bass, or low-frequency, response may give a stronger rhythm to dance to. The circuits which do this are called "tone controls".

The treble control normally has a mid position at which it has no effect. Rotating the control clockwise from the mid position increases the gain of the pre-amplifier to high frequencies; rotating anticlockwise reduces the gain at high frequencies. The control usually affects the response most at frequencies from about 10 kHz to 20 kHz, and progressively less down to about 1 kHz.

The bass, or low-frequency, control works at frequencies below 1 kHz, its greatest effect being at about 100 Hz, and progressively less up to 1 kHz. Rotating the bass control clockwise from the mid position increases the gain of the preamplifier at low frequencies; rotating anticlockwise decreases the gain at low frequencies. Typical gain curves for tone controls are shown in Fig. 16.

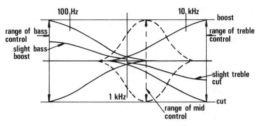

FIG. 16. Tone controls

Sometimes an amplifier has a mid-frequency control which operates at a frequency a little above 1 kHz. It may only boost, or both boost and cut, that frequency and those near it – mid-frequency cut never occurs on its own. A typical mid-frequency response curve is shown dotted in Fig. 16 for a version with both boost and cut. The effect of this control is sometimes known as

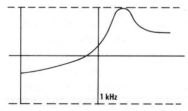

FIG. 17. Frequency response produced by slight bass cut, full mid boost and slight treble boost

The Amplifier

"presence", as it seems to liven up music and bring it closer to you in the room, albeit rather artificially.

As an example of the use of tone controls, consider applying slight bass cut, slight treble boost and a lot of mid frequency to a preamplifier. A response such as that shown in Fig. 17 would be obtained. This would sound brassy but bright, owing to the treble and mid-frequency boost, but would feel slightly distant because of the lack of bass, this being a real effect of distance, when bass notes are scattered as opposed to the artificial effect produced by use of presence. This arrangement would, however, liven up a bass heavy but otherwise dead recording.

Ideally a preamplifier should also have low- and high-frequency filters on it. The low filter reduces low frequencies and hence effects such as rumble and mains hum, and so is particularly useful when listening to records. It will not in fact always be the record player which is introducing these effects. Even some high-quality records have rumble and hum recorded on them, particularly historical recordings made before recording techniques were as sophisticated as they are today.

The high filter removes very high frequencies and will lessen the effect of scratches on a record, or hiss on a poorly received radio station. It, too, may be useful when listening to a historical recording where, for instance, a "78" record, which is often hissy, has been reproduced on an LP to make a particular performance available.

Filters also have a property called "slope". This is an expression of how severely they reduce signals beyond their operating frequency. The simplest reduce at a rate of 6 decibels for each octave away from this, *i.e.* 6 dB/8ve. The next stage of complexity is 12 dB/8ve. Ideally if steeper filters than this are fitted they should not be just switched in and out since their effect is so severe that slightly less might be required. Thus a knob should be

The Amplifier

provided to adjust the effect of filters above 18 dB/8ve and their frequency should be selectable as well. 6 dB/8ve is all right for a low filter but barely adequate for a high filter, where faults are likely to be more severe. The steeper the filter is, and the more flexible, so it may better be used to remove only those signals intended.

VOLUME AND LOUDNESS CONTROLS

The volume control of a preamplifier controls the level of the signal which is fed to the power amplifier and hence how loud it is from the loudspeakers, since the power amplifier has a fixed gain. It should be noted that, at low volumes, the ear is less sensitive to high and low frequencies than to mid frequencies. The treble and bass controls can be used to correct for this by applying a small amount of boost to each when the volume control is turned down. Sometimes a switch, called a "loudness control", is provided, which applies this boost when in the "on" position. Alternatively, but rarely, the preamplifier automatically provides the necessary boost when the volume control is turned down. This automatic boost within the preamplifier is not necessarily a good thing since it will work at a fixed volume setting so that the volume at which it acts will depend on the efficiency of the power amplifier and loudspeakers and will vary from system to system.

BALANCE CONTROL

The balance control varies the gain of one channel of a stereo system with respect to the other while maintaining the overall power output. Although, ideally, a stereo system should be perfectly matched, so that the sound produced by each loudspeaker is the same, it may not be possible to have loudspeakers symmetrically placed in a room or for them to be equidistant from the listening position.

The Amplifier

A loudspeaker placed in a corner will be more efficient than one in the centre of a wall since the corner acts like a megaphone making that speaker louder. The balance control is designed to compensate in part for this, though it cannot fully compensate as there will probably be tonal differences as well as those of volume.

Similarly, if one's favourite chair is nearer to one speaker than the other, the nearer loudspeaker can be turned down with the balance control. Again this is only a partial correction as the brain interprets sound reaching the ear as coming from the sound source first heard and, even though the distances may be very small and the volume of sound from each speaker may seem the same, the sound from the nearer speaker will reach the ear first. So a central signal will still appear to come from the closer speaker.

STEREO/MONO SWITCH

The stereo/mono switch on a preamplifier performs the function of keeping the signal paths separate in the stereo mode, or joining them so that the same signal is fed to both loudspeakers in the mono mode. Stereo records would probably be listened to with the switch in its stereo position. Mono records should always be played with the switch in the mono position as sometimes the quality of the reproduction may be reduced due to the pick-up of accidental vertical modulation of the record groove. This is antiphase signal and hence cancels when the channels are summed. On a mono cartridge it would not be there as the cartridge does not pick up this motion.

THE POWER AMPLIFIER

The preamplifier has taken various different types of input signal and converted them, where necessary, to flat signals. It has also enabled the listener to make adjustments to the tonal quality of the programme material

The Amplifier

to suit his own tastes and to correct imperfections in the sound system or the programme material itself. The power amplifier now needs only one type of input, that is one suitable to be driven by the output of a pre-amplifier; and it has only one function, that is to reproduce the signal fed to its input with as little alteration in its content as possible but at a much higher power.

Unfortunately most power amplifiers produce some distortion, that is alteration in signal quality, and it can sound most unpleasant. Distortion is always a function of the frequencies present in the signal and, though it is related to them by the laws of physics, it may not be by those of music. If it is not, then it jars on the ears and is very noticeable. Unfortunately it is easier to get rid of the musically related distortion when designing an amplifier than to get rid of the other kind, though neither should be there. An acceptable figure for distortion is less than 0.1 per cent as a general rule-of-thumb, though higher figures can be tolerated as long as they are the musically related kind. Two amplifiers which have the same measured distortion figures may sound very different, so the final test is to listen to them.

The power-handling capacity of a power amplifier is also of interest, though one rarely listens to sound at much greater than 3 watts average value (and that is loud!), which will be well under the power rating specified. However, the power rating also expresses how great an instantaneous or peak voltage the amplifier will provide. Now, though the average power value of a piano note, for example, is quite low, it will have quite a large volume for a very short time at the very beginning when the hammer hits the string; that is it will produce a large peak voltage. If the amplifier cannot handle this peak voltage it will distort quite noticeably with a characteristic buzz. The peak voltage produced by the piano note being struck may require a power rating of

The Amplifier

20 watts for one-fiftieth of a second, though the rest of the note may only be at an average power of 2 watts.

For this reason a power amplifier should have as high a power rating as possible, commensurate with the power-handling capability of the loudspeakers it is to drive. A reasonable minimum figure is 15 watts, though over 20 watts would be preferable.

Summary

1. Check the amplifier chosen has suitable inputs for all the sources you are likely to feed into it.
2. Check that you will be able to master the use of the controls – there is little point in paying extra money for controls that are never used.
3. Check the power available is adequate and suitable for the speakers chosen. Preferably both should be able to handle greater than 15 watts r.m.s.

6 | THE LOUDSPEAKER

So far we have seen that acoustic waves in air are converted into electrical waves so that they may be more easily recorded for later playback, or transmitted over long distances, and we have discussed the equipment which the listener has for recovering the recorded or transmitted waves as exactly as possible in an electrical form. These electrical waves must now be converted back to acoustic waves for the ear to receive. The loudspeaker performs this last function.

RESONANCES

Of all the elements of a hi-fi system, the loudspeaker is probably the most difficult to specify, and yet it is critically important to the sound quality of the complete system. The performance of a speaker depends to a certain extent on its surroundings, so absolute measurements will not really tell the prospective buyer which one sounds best. In general, though, the flatter the response the better (and the more expensive), though the ear will tolerate a fair amount of variation in response throughout the audio range.

All mechanical systems, which speakers are, tend to resonate at particular frequencies which are a property of the particular system structure. This means that it is easier for the speaker to radiate sound at its resonant

The Loudspeaker

frequencies than at all others, and so it sounds louder at those frequencies than at others. This gives peaks in the response. The converse is also true, giving dips in the response at certain frequencies.

Dominant resonances should be avoided as these will give programme material a distinctly "coloured" sound which can be unpleasant. Slight resonances may make the programme material sound brighter for the same reason given in the last chapter when describing the effect of mid-frequency, or presence, boost. A speaker with slight resonances may therefore be more pleasing to the listener than one without any detectable resonances at all.

When choosing a loudspeaker, a listening test is essential, and this should be done using the same type of amplifier which has been chosen for the final system. Differences in the drive conditions given by different types of amplifier can make the loudspeaker sound very different. It may not happen, but by making sure you may prevent an expensive disappointment.

DRIVERS AND CROSSOVER CIRCUITS

In order to reduce the likelihood of resonances, loudspeakers can be designed to operate over less than the audio range. This means that, in certain cases, two, three or even more speakers may be combined to cover the full audio range, all of them usually being mounted in one cabinet.

The bass unit (affectionately known as the "woofer" because of the sort of sound it produces, though this term is tending to be less used) may handle frequencies up to 3 kHz in a two-speaker system. The treble unit (called the "tweeter") would handle frequencies above this. Alternatively the audio range could be divided into three, at a frequency somewhere in the region of 300 Hz to 600 Hz and again at about 5 kHz, a mid-range unit

The Loudspeaker

being used to handle the frequencies between the bass unit below and the treble unit above. Dividing the full audio range between different specialist speakers tends to even out the response of the total loudspeaker set, though there is a law of diminishing returns in the number of units used, a practical limit being four. The individual units are also known as "drivers", *i.e.* bass driver, mid driver, etc.

Circuits called "crossovers" filter the signals reaching each driver so that they do not receive those parts of the audio signal which are outside their operating range. This prevents erratic response due to, for instance, unavoidable resonances in the drivers, since these will be designed to be outside their range wherever possible.

The crossover circuits used in a particular loudspeaker will be designed for that particular combination of drivers in that specific enclosure and the performance of the loudspeaker will almost certainly be degraded if any of these are changed. When making loudspeakers from a kit, therefore, it is essential that the design is rigidly adhered to.

The bass driver of a set of speakers will be the largest and will have the softest suspension. This is because the displacement of a speaker to produce a given volume of sound is inversely proportional to the frequency. Hence the bass speaker has to displace the furthest. It can often be seen to move, especially for loud, low notes. In addition, the efficiency of a speaker is in part dependent on the size of the radiating area compared to the frequency of the sound radiated; the lower the frequency the larger should be the radiating area. The bass driver will therefore have the largest diameter.

The same considerations hold for the other drivers. The mid driver will be smaller than the bass driver since it handles higher frequencies and therefore requires a smaller radiating area and displacement. The treble

The Loudspeaker

driver will be the smallest and may be what is known as a "dome tweeter", or merely a smaller version of the other drivers. Never touch the front of a treble driver; it has to move very fast indeed to follow high frequencies and hence its moving parts are very light and fragile. It is not wise in fact to touch the moving parts of any speaker as a general rule, but at least the lower-frequency drivers are less delicate and vulnerable to damage from accidental touching.

Fig. 18(*a*) shows a typical cross-section of a driver unit. Fig. 18(*b*) shows the cross-section of a dome tweeter.

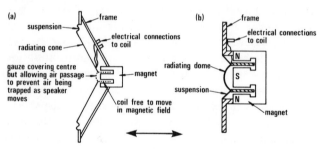

FIG. 18. Cross-section of speaker drive units

The coil carries the drive current from the power amplifier and is situated in a magnet. The action of the speaker depends on the fact that a wire carrying current in a magnetic field experiences a force which tends to move it. The magnetic field is arranged in such a way that a current in one direction moves the coil forwards and a current in the other direction moves it backwards, the movements being in the direction of the arrow shown in Fig. 18. The coil can move the same distance in both directions from the mid position at which it rests when carrying no current. The force on the coil generated by the current it carries is balanced by the pull of the suspension, so that a given current produces a given displacement. The loudspeaker cone or dome is attached to the coil and moves with it. Hence the displacement of the radiating surface of the driver follows the instantaneous currents in the coil which are defined by the signals supplied by the power amplifier.

ELLIPTICAL SPEAKERS

It was said that the efficiency of a speaker is a function of its size and so, in fact, are its resonances. Because of this, elliptical speakers are sometimes used, since these effectively combine several sizes in the same unit. The long axis defines the performance at low frequencies, though, because of the reduced radiating area, they are less efficient than a conical speaker of the same diameter as the long axis. Conversely, the short axis defines the performance at high frequencies, though again it is less efficient than an equivalent cone owing to the larger mass which has to move. There are, however, an infinite number of sizes in between and the result is that, rather than having a small number of dominant resonances, an average response is obtained which is much flatter than it would be for a similar quality of speaker with a circular cone.

Another technique is sometimes used to effect a reduction in size of a speaker cone as the frequency rises. The cone is constructed in such a way that for low frequencies it moves as a unit, but, as the frequency rises, the connection between the edge of the cone and the centre where it is driven becomes less and less rigid, so that at high frequencies only the very centre of the cone is being driven. This is the reason for the concentric rings which are sometimes embossed in the cone, though with modern materials the cone may be manufactured from plastics which display this characteristic without complicated moulding shapes.

Sometimes, also, a technique similar to that described

The Loudspeaker

above is used where a large and a small cone are mounted separately on the same drive coil. Above a certain frequency the large speaker is ineffectively driven and the small speaker takes over. In this arrangement, the small speaker is known as a "parasitic speaker" and the whole assembly is a "twin cone speaker".

LOUDSPEAKER CABINETS

There are basically four main types of loudspeaker cabinet in which a driver can be mounted. These all affect the performance of the speaker, and the dimensions of them are critical. The outside shape of the box is more or less immaterial; this is merely the case in which the working parts are mounted.

The most common form of small loudspeaker cabinet is called the "infinite baffle". The reason for this is that sound coming from the front of the speaker is separated from sound coming from the back of the speaker by a screen which in effect stretches to infinity in all directions in the plane of the diameter of the cone. For a given signal which, say, moves the cone towards an observer in front of the speaker, an observer behind the speaker will see the cone move away from him. If these two signals were both to reach the listener, with a difference in path lengths between the signal from the front of the speaker and the signal from the back, as in Fig. 19(*a*), then for certain wavelengths the signals would reinforce one another but for others they would destructively interfere.

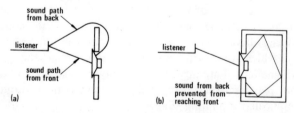

FIG. 19. Baffle and infinite baffle mountings

The Loudspeaker

This would result in a very uneven response. The infinite screen, usually in the form of a box which encloses the back of the speaker as shown in Fig. 19(*b*), prevents this from happening.

Care must be taken when designing the enclosing box since, being a sealed cavity, it will tend to resonate. Usually the inside of the box is lined with sound-absorbent material to dampen potential resonances. It is also important that the box is airtight, since the drivers used in these speakers are designed to be used in airtight conditions and will not give a flat response otherwise. Infinite baffle cabinets have only medium efficiency, but they are small and reasonably inexpensive.

A bass reflex cabinet is the converse of the infinite baffle and uses the signals from both the front and back of the driver. A driver may have a good flat response above its natural resonance, but below it the response falls off as the frequency decreases. In the bass reflex cabinet, a controlled difference in sound path lengths is allowed between the signals from the front and back of the driver and the main resonance of the combined signals is designed to happen below the natural resonance of the driver, hence reinforcing its low-frequency response where it would otherwise be falling off. In addition, the signals destructively interfere at the natural resonant frequency so that the peak in response which would otherwise occur at that frequency tends to be cancelled. Bass reflex cabinets have medium efficiency but extended bass response.

Another type of loudspeaker cabinet is the horn. Horn-loaded loudspeakers are designed to raise the efficiency of the drivers. Ideally a loudspeaker should be of a size approaching the cross-sectional area of the room it is being used in for maximum efficiency. Not only would this be impractical – the sheer size of a cabinet to hold such a driver would be imposing to say the least –

The Loudspeaker

but also it just would not work, especially at high frequencies when the mass would be subjected to enormous acceleration requiring vast amounts of power.

The compromise solution, which still results in rather large loudspeakers, is to attach a tube to the front of the driver, the attached end of the tube being the same size as the driver while the other end is considerably larger. This tube is known as the "horn" and is illustrated in cross-section in Fig. 20(*a*). The precise shape and sound-path length of the horn is critical to the response of the speaker. Sometimes the horn is folded back on itself as shown in Fig. 20(*b*) to reduce its overall length.

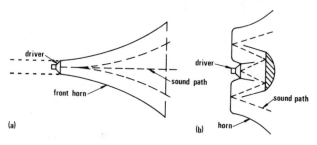

FIG. 20. Horn mounting

The computations involved in designing the shape of the horn are complex and, in any case, there is some disagreement as to the optimum shape. The effect produced, however, is that of a megaphone, though the quality is quite high. The same system in reverse may be used to load the back of the driver so that the signal from the back is effectively lost. This is shown in the dotted lines in Fig. 20(*a*) above and it is clear that the tube must still be rather long to effect a significant reduction in the signal, leading to an even larger cabinet.

A transmission line loudspeaker is designed to overcome another loading effect on drivers which tends to introduce distortion. If a driver does not have the same freedom of movement in air both in front and behind it, then it will move more easily in one direction than in the other and so will distort the signal presented to it. A complex combination of a tube and sound-absorbing materials is therefore fitted behind the driver so that the loading of an average room is simulated within the speaker enclosure, thus creating the same environment in front and behind it. This results in a very inefficient loudspeaker, but with lower distortion than other kinds. The name "transmission line" is borrowed from an engineering term describing a similar electronic effect.

ELECTROSTATIC LOUDSPEAKERS

A different type of loudspeaker altogether is the electrostatic type. In this the cone is replaced by a very light diaphragm which is charged with a high d.c. voltage relative to a nearby perforated plate. When the audio signal is applied to the diaphragm, it causes fluctuations in the charging voltage and hence the force on the diaphragm so that the diaphragm moves with respect to the perforated plate as the signal changes. The speaker is constructed as a sandwich with the diaphragm placed in fact between two perforated plates so that uneven effects due to the diaphragm approaching one plate are offset by it moving away from the other. In this way distortion is kept to a minimum. The sound is radiated from the diaphragm through the perforations in the plates.

Electrostatic loudspeakers produce crisp, clean sounds in the treble and mid-frequency ranges, but must be used at a reasonable distance from surrounding walls – at least 6 feet is recommended – to achieve a good response at bass frequencies. In order to overcome this problem while still taking advantage of the clean response at higher frequencies, electrostatic treble and mid-frequency units are sometimes combined with a conventional bass driver. However, since a totally electrostatic loudspeaker

The Loudspeaker

is very thin – it is not much thicker than a central heating radiator for instance – it can be used where space is at a premium if the loss of bass is acceptable.

The various types of driver and cabinets we have discussed above are frequently combined together in one loudspeaker in order to take advantage of the best features of each for the different frequency ranges. For example, a combination which is used by one manufacturer is:

>   Bass: Transmission Line
>   Mid: Reflex
>   Treble: Infinite Baffle
> High Treble: Infinite Baffle

PHASE LINEARITY

A concept that has been taking more and more designers' time recently is phase linearity. In this, not only do the relative amplitudes (*see* Chapter 10) of the components of a signal remain the same through the full frequency range, but also they have exactly the same timing with respect to each other as was originally recorded. This is rather more difficult to achieve. Although on a mono signal the difference is negligible, with stereo the ear uses this timing information to identify precise positioning of the signal source. Phase linear speakers therefore give a more realistic sound image than do those with merely a linear amplitude characteristic. There is controversy about this, but the fact remains that most of the more popular speakers designed before much attention was paid to phase do have good phase linearity.

GENERAL CONSIDERATIONS

The size of a loudspeaker gives some idea of its performance. Normally larger loudspeaker cabinets indicate a higher power-handling capability, and the loudspeakers

The Loudspeaker

chosen should be capable of handling all the power that the amplifier is able to supply. Note that the amount of power an amplifier gives is dependent to a certain extent on the impedance of the loudspeaker it is driving; refer to Chapter 1 for a discussion of this, however.

For speakers of a given power rating, the larger the cabinet, the better the response to very low frequencies as a general rule. Although the loudspeaker will produce lower frequencies as the cabinet gets larger, however, the quality of this bass response will depend on how massive the construction is; the more solid the better the quality, regardless of size. Bear in mind, however, that it is pointless to have a loudspeaker with a bass response down to 20 Hz in a small room. The room itself is part of your listening apparatus and the cut-off frequency of the room is inversely proportional to its length. A room 17 feet long has a cut-off frequency of approximately 60 Hz; to hear sounds at 30 Hz, it must be 34 feet long! The proportions of the room are also important. It should be a bit longer than it is wide and a bit wider than it is high for optimum effect, with the loudspeakers set across the short wall. Low ceilings tend to produce a rather hollow sound.

The final point to be made in this section is that high-frequency drivers are very directional; that is, the best response is obtained by listening straight in front of them. To the sides, above and below, they are less audible. This is why they are mounted near the top of loudspeaker cabinets so that they are as near as possible to the correct height for the seated listener when the loudspeaker sits on the floor. Loudspeakers should not, therefore, be used upside down, even though this is often more convenient for the routing of wires back to the amplifier. Also, if a small loudspeaker is used on its side, on a bookshelf for instance, care should be taken that it does not point significantly away from the listener. Trial and error using

The Loudspeaker

a record with a lot of high frequency is the best way to check this.

HEADPHONES

Using headphones is a convenient way of listening to music or other programme material with minimum disturbance to other people. They also give an effect when used with a stereo recording which many people like, the sound seeming to come from on top of the head, but which may be a little disturbing when one is not used to it. They are also essential when making live tape recordings.

The range of headphones on the market, as with loudspeakers, is enormous, ranging from cheap and cheerful to electrostatic and expensive. They are all in effect a pair of very small loudspeakers mounted at either end of a band so that they can be held on the head and positioned over the ears. It is well worth spending at least £15 on a pair of headphones if they are going to be used to any reasonable extent, since comfort, low distortion and low resonance start at that price in general. It is just as important that they are comfortable to wear as that they produce good-quality sound since, if they are used at all, they are likely to be worn for fairly long periods. Once again a listening test is definitive. Care should also be taken that the headphones have an impedance suitable for the amplifier they are to be used with.

Summary

1. Check the loudspeaker can handle the power you intend to feed it; if not, it may be damaged rather expensively.
2. Check that the loudspeakers chosen are suitable for the room they are to be used in both for performance and as furniture.
3. Headphones are often a useful addition, but check that

you like using them; some people do not. Cheap headphones are good for the experience, but are rarely high fidelity. Listen to them and make the choice after becoming used to using them.

7 | THE TAPE RECORDER

All domestic tape recorders operate on the same principle, although their mechanics vary in detail. A plastic-coated tape, coated on one side with a magnetic oxide akin to rust, is passed across small electromagnetic heads which either force a magnetic pattern on to the tape or pick up a pattern already on the tape without disturbing it.

TYPES OF TAPE MACHINE

The three types of tape machine – open reel, cassette and cartridge – have differences in the way the tape is stored, in the arrangement of the tracks on the tape and in the dimensions of the tape itself. One type of tape machine will not be able to play the tapes designed for another, although a cassette player/recorder may be built in the same box as an open-reel one to allow both types to be used.

An open-reel machine has two separate reels, each of which is designed to carry $\frac{1}{4}$ inch wide tape. The tape is initially fully wound on the left-hand reel, the right-hand reel being empty. Playing the tape causes it to unwind from the left-hand reel, pass across the heads and wind on to the right-hand reel. The disadvantages of this system are that the user has large reels to store, and the tape has to be threaded through the machine on to the right-hand reel by hand to set it up before it can be used, which can be a

The Tape Recorder

nuisance. The advantages are that the reproduction quality is the highest of the three systems, and editing the tape is comparatively easy, it being practically impossible for cassettes and cartridges. On an open-reel tape, up to four separate tracks can be recorded normally with a reasonable degree of fidelity, giving four mono tracks or two stereo ones as a usual maximum.

A cassette machine carries a cassette, which is a small, flat box with two small reels built into it and a small slot in the narrow side through which the tape is accessible. The reels carry $\frac{1}{8}$ inch wide tape which is wound on the reels inside the cassette and sealed in the factory. As with open-reel machines, the tape passes from the left-hand reel, across the heads and on to the right-hand reel. The tape is automatically positioned in the machine when the cassette is inserted, so the awkward job of threading the machine is avoided. The other advantage is that the cassettes are much smaller than open reels and are easier to store and less prone to damage. The disadvantages of this system are that the sound quality reproduced has been lower, although modern improvements are overcoming this, and editing is very difficult as the tape is very delicate and easily damaged outside the cassette.

A cartridge machine is similar to a cassette machine in that the tape is pre-loaded into the cartridge at the factory. The tape used is $\frac{1}{4}$ inch wide and the arrangement of the tape on the reels is different so that a cartridge is a different shape to a cassette and is about three times bigger in size. The advantages and disadvantages of a cartridge machine are similar to those of a cassette, with the additional advantage that a cartridge is simply pushed into the machine whereas a cassette is clipped in at an angle and pushed flat, though some cassette machines have been designed to do away with this double-action insertion. However, in general, cartridges are more easily used in a car, say, where the driver will need the simplest

The Tape Recorder

loading method available in order not to distract him from his driving. Another difference is that cartridges are designed so that they will play continuously, repeating whatever is recorded on them, whereas a cassette has an end to its playing time and must be turned over or rewound to continue playing.

OPERATION OF A TAPE RECORDER

A tape machine can have two or three heads. In describing the operation of a tape machine, a three-head machine will be considered first, since the action of a two-head machine is most simply described in the light of this.

The three heads arranged as shown in Fig. 21 are the erase head (E) which cleans the tape of any signals previously recorded on it, the record head (R) which puts a new magnetic signal on the tape, and the playback head (P) which detects these signals and reproduces them

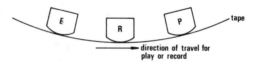

FIG. 21. Three-head layout

electrically without disturbing them. The tape passes over these three heads in succession during the recording and playback processes. For simple playback, the erase and record heads are turned off so that the signal recorded previously on the tape is picked up by the playback head and not destroyed by the other two.

The three heads are built similarly with only slight changes in detail. A simplified sketch and cross-sectional diagram of the main features are shown in Fig. 22. The erase head has a wide gap, perhaps 0.5 mm across, filled with a plug made of a non-magnetic material. The magnetic field produced in the gap by electrical

The Tape Recorder

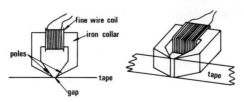

FIG. 22. Schematic of tape heads

signals in the coil tends to take the easy path through the tape to pass from one pole to the other, rather than pass through the non-magnetic plug. This magnetises the tape. Now the electrical signal applied to the coil on the erase head is at a frequency higher than the audio range and is stronger than audio signals applied to the record head. The magnetism put on the tape as it passes across the erase head therefore swamps the magnetism due to audio signals already on the tape. The erase signal has gone through several positive and negative cycles as a particular piece of tape passes the gap, due to the high frequency of the signal and the wide gap. The result is that nothing is effectively recorded on the tape. Any residual magnetism which might be left on the tape produces a signal out of the audio range when played back anyway, so the tape is effectively cleaned.

The record head has a much smaller gap, approximately 0.02 mm, and this is sometimes filled with a non-magnetic plug, though not always. The coil in this head carries the audio signal to be recorded and creates a fluctuating magnetic field in the gap which is imposed on the tape as it passes across the head. The tape passes the head at a constant speed so that a given number of magnetic cycles per second from the head puts a given number of cycles per given length of tape. The gap in the head must be small enough so that, even at high frequencies, the tape has moved the full width of the gap before the instantaneous signal in the head changes noticeably. If

The Tape Recorder

the gap were larger than this, a blurring effect would be produced by the interaction of new signals on the signal just recorded and high frequencies would be lost. An alternative to using a very small gap for good high-frequency response is to run the tape at a higher speed. This can be expensive on tape but does, however, give the added advantage of reducing any small imperfections which might be on the tape. The gap in the record head must nevertheless be of a reasonable size or else the magnetic field produced by it will not be large enough to penetrate far into the tape, and the signal recorded will be weak. Also, it is the trailing edge of the magnetic field which actually leaves the signal on the tape, *i.e.* the tape is magnetised as it passes out of the gap rather than in the middle, so extremely fine gaps are not necessary.

The playback head picks up the magnetic field on the tape. This field radiates out from the tape slightly and the head causes it to pass through the coil, thus generating a small electrical signal. In this case there is no non-magnetic plug in the head gap, since the field radiated from the tape must be encouraged to pass between the poles in order to pick up the signal. For the same reason, the gap must be very small indeed, approximately 0.005 mm. Also, if the gap is too large, high frequencies will be lost since both the positive and negative cycles of the signal on the tape will be within the width of the gap at the same time and will add together to produce nothing.

The operation of a two-head machine is similar, except that the functions of the record and playback heads are combined in one head, with associated electronic circuitry being switched to accomplish whichever is required. Naturally this is a compromise solution which trades off quality against cost. Apart from giving better quality, the three-head machine has a further advantage in that the playback head can be monitored during the making of a recording. In this way the quality of the recording itself

The Tape Recorder

(and not merely the signal prior to recording) can be checked at the time it is made.

Cartridge recorders are invariably two-head machines. There are some three-head cassette recorders on the market at the time of writing, but they are rather expensive and the two-head type is normal. Cheap open-reel machines tend to be two-head, but those with any legitimate claim to high fidelity will be three-head.

The arrangement of tracks on the tape varies a lot, though it will be consistent for any one type of machine. The normal arrangements for playback will be discussed here and obviously the formats for record will be identical. The tape may be considered to be divided into strips along its length, each strip being a track and being the width of the tape affected by the tape head. Thus a full-track head operates over the full width of the tape, whereas a half-track head operates on only half of the tape so that the tape must be turned over to use the other half. This is illustrated in Fig. 23. A small strip of tape is always left unused at the edges of a track to allow for mechanical errors and so that adjacent tracks do not interfere with each other. The tracks are not visible on the tape, though the arrangement of the heads can be seen on close inspection of the machine.

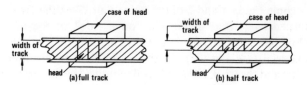

FIG. 23. Layout of tracks for full-track and half-track machines

In a cartridge machine, the tape has eight tracks and plays only in one direction. The first time the tape passes the head, tracks 1 and 2 are picked up. When the end of

The Tape Recorder

the tape is reached, the head moves down to pick up tracks 3 and 4. The start of the tape is joined to the end to create a continuous loop which is wound in an ingenious way inside the cartridge. The machine carries on playing when the end of the tape is reached with only a slight pause while the head is repositioned by the mechanics. This process is repeated when the end of the tape is reached a second time so that tracks 5 and 6 and then tracks 7 and 8 are picked up. This is illustrated in Fig. 24.

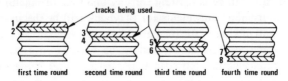

FIG. 24. Layout of tracks in a cartridge machine

In the quadrophonic (*i.e.* four-channel) mode, as opposed to the stereo mode, the tape travels round twice instead of four times for a complete cycle: tracks 1, 2, 3 and 4 are picked up on the first time round and tracks 5, 6, 7 and 8 are picked up the second time round.

On a cassette machine there are four tracks available on the tape. Tracks 1 and 2, counting from the top, are a stereo pair played in one direction, and tracks 4 and 3 become tracks 1 and 2 of the stereo pair when the tape is turned over, thus giving the "other side" of the tape. Only one side of a tape is actually coated in magnetisable material (*see* Fig. 25). In this case the tape heads are fixed and the tracks selected are determined by the position of the tape itself.

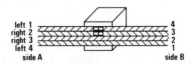

FIG. 25. Layout of tracks for a stereo cassette

The Tape Recorder

If the cassette machine is a mono one, then it is a half-track machine, and if the tape is in stereo the head picks up tracks 1 and 2 simultaneously, adding them together. Thus a cassette stereo system is mono compatible without special equipment. A mono-recorded cassette tape, which will be half track, will play back in mono on a stereo machine.

An open-reel machine may be full track, half track or quarter track. A full-track machine is as shown in Fig. 23 (*a*). Since there is only one track available, it will record only in mono and play only in one direction, the sounds recorded being played back in reverse if the tape is turned over. Half-track mono is as shown in Fig. 23 (*b*), track 2 being the "other side" of the tape when it is turned over. Half-track stereo uses half a track for each channel as shown in Fig. 26, the sounds again being played back in reverse if the tape is turned over.

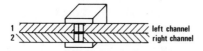

FIG. 26. Layout of tracks for half-track stereo

Quarter-track mono is arranged so that tracks 1 and 3 play back in one direction, and tracks 2 and 4 play back in the other direction, a switch on the machine selecting one of the two tracks available on each side. This switch is usually marked 1:4 and 2:3, so any one of the four tracks may be used by having the tape the right way up and selecting the right head (*see* Fig. 27).

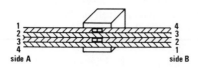

FIG. 27. Layout of tracks for quarter-track stereo

The Tape Recorder

Quarter-track stereo is similar except that both tracks 1 and 3 or 4 and 2 are played simultaneously to give the stereo channels; *i.e.* one stereo track is available on each side of the tape. By separating the tracks used for a stereo pair by a track on the other side in this way, interference between channels within the head assembly is reduced to a minimum. This does mean, however, that a half-track mono tape cannot be played on a quarter-track stereo machine, as the lower head will pick up the other side of the mono tape in reverse. It would be possible of course if the other side of the mono tape were left clean.

Quarter-track machines use all four track positions available to them; *i.e.* they have four heads in the head assembly. The tape is recorded in one direction only, any one or any combination of heads being selectable by a switch on the machine so that it can be used in mono, stereo or quadrophonic mode. Alternatively, a combination of heads can be selected so that the tape can be made to conform to one of the other standards.

TRACK WIDTH AND TAPE SPEED CONSIDERATIONS

Although the record head records very well over most of the track, the very edges of the track tend to be under-recorded. To compensate for this effect, the widest track width possible should be used so that the edge effects are insignificant compared to the full track width. It obviously takes twice as much tape to record something in half-track stereo rather than quarter-track stereo, and so is twice the cost. It is therefore up to the user to decide how he wishes to trade off quality against expense.

Another tape consideration is the noise introduced by the grainy surface of the tape due to the magnetic coating being a powder deposited on the plastic backing, even though it looks smooth. The noise introduced is less noticeable when the tape is run at high speeds, and the

The Tape Recorder

frequency response of the recording is also better, but again more tape would be used for a given recording at high speed than at low speed. So the user must also decide on the speed to use for his requirements. Most open-reel recorders have two or three speeds available: $1\frac{7}{8}$ inches (4.76 mm) per second is a very economical speed but will obviously give the poorest reproduction; $3\frac{3}{4}$ inches (9.5 mm) per second is a reasonably economical speed giving reasonable quality; $7\frac{1}{2}$ inches (19 mm) per second gives good fidelity on modern machines. An even greater speed, 15 inches (38 mm) per second is available on some machines. This gives very high quality and is the modern studio standard.

There is no choice of speed on cassette or cartridge machines.

TAPE DRIVE MECHANISMS

The tape is pulled past the heads by a drive system consisting of a capstan and a pinch wheel (*see* Fig. 28).

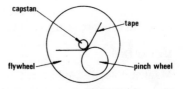

FIG. 28. Schematic of tape drive mechanism

The capstan is an accurately ground shaft rotating at a fixed speed with a flywheel, or sometimes electronic circuitry, to ensure smooth rotation. The pinch wheel is usually made of rubber and presses the tape against the capstan so that it does not slip. In this way the tape is pulled past the tape heads at constant speed. The pinch wheel only holds the tape against the capstan during record and playback. For fast wind and rewind there are motors under the reels which pull the tape, and the capstan does not regulate the tape speed. These reel motors serve to keep the tape tight, taking up any slack in it, during record and playback.

Sometimes only one motor is used for all three functions, *i.e.* the capstan and the two reels, and an arrangement of drive belts and clutches is used to change its mode of operation. This system is cheaper but, for good performance and reliability, it is better to have three separate motors.

In cassette machines, twin capstans are sometimes used because of friction problems due to the parts being smaller and the mechanics generally more difficult. In this system, one capstan operates before the tape reaches the heads, and one operates after the heads. The first one provides a back tension on the tape by running slower but with a slipping drive mechanism. The one after the heads governs the speed of the tape. In this way, the effect of friction within the cassette acting on the reels is isolated from the part of the tape actually in contact with the heads. The improvement in performance of a twin capstan cassette machine over a single capstan one is just audible, and again it is up to the user to decide whether the extra expense of the twin capstan machine is worth it.

METERING

Magnetic tapes can accept only a certain amount of magnetism, and so a tape recorder will have a meter, or meters, on it to check that the signal fed to the tape is not so large as to saturate it and cause distortion. The meter generally has a red section marked on the right-hand end of the scale and there will be distortion on the tape if the needle is consistently in this section.

The recorded signal must be as large as possible without distortion, however, to overcome noise generated by the tape. The tape must pass very close to the head, a matter of thousandths of an inch, or else high frequencies in particular will be lost. However, the granu-

The Tape Recorder

lar surface of the tape causes minute random signals to be generated in the playback head as it passes across the head. The recorded level must be considerably greater than the noise generated by the tape for good-quality recording, and so, as a rough guide, the recording level should be set to keep the needle flickering just below the red section of the meter, though occasional flicks into the red region are allowable. Further discussion of the use of the recording level meter is included in Chapter 11.

TAPE RECORDER CONTROLS

The tape recorder will have a variety of controls depending on its complexity. These fall into three groups: mechanical, input and output.

The mechanical controls are as follows: fast wind and fast rewind, for moving from one part of the tape to another at a much higher speed than the playing speed – the tape is not held close to the heads during these functions to avoid unnecessary wear; play, for positioning the tape close to the heads and running the tape at a constant speed past the heads; record, for performing similar functions to the play control but for recording purposes; and stop, for stopping the tape moving and releasing it from close contact with the heads.

When the record control is used, the erase head automatically comes into action to clean the tape. The record control is usually interlocked with the play control so that it cannot be used after the play control has been pressed, thus preventing accidental erasure of the tape. For the same reason, the record control must usually be operated at the same time as the play control in order to get into the record mode, so that it is difficult to do this accidentally.

The mechanical controls, with the exception of the record control, may be combined into a single lever, since only one of them would be used at any one time.

The Tape Recorder

The input controls may be as follows: an input selector, for a machine which has a variety of input sockets and sensitivities; recording level controls, to adjust for variations between different types of input from the same source, *e.g.* classical music, speech, pop music, etc.; a stereo/mono switch of some form; and a track selector on machines with more than one track.

The output controls will be: a volume control, to control the playback volume independently of the recording level; a balance control on a stereo machine; a track selector, though this may be combined with the input track selector; and tone controls, though these are normally fitted only to machines with their own power amplifiers – a tape deck only without an amplifier of its own would be used with an amplifier which would have its own tone controls.

In the case of a tape machine with its own power amplifiers, small loudspeakers may be fitted inside the cabinet or in the lid, making it a complete unit. The quality of these loudspeakers is, however, likely to be lower than that of separate ones, so output sockets for the connection of alternative loudspeakers should be provided.

TAPES

The choice of tape for a tape recorder is important, since the machine will be adjusted for a particular tape. A simple rule-of-thumb is to buy tape of the same brand name as the tape machine, if this is available. This is frequently, in fact, tape produced by another company with the name of the tape recorder manufacturer on it. Check with your supplier if you have difficulty obtaining a particular brand.

Alternatively, the "get what you pay for" rule holds quite well, but experiments with different brands are best to find the tape most suited to the machine. Using the

The Tape Recorder

wrong brand of tape, even though it is a high-quality one, can sometimes give higher distortion levels and more noise or background hiss on a recording than one might reasonably expect.

Cheap tapes are to be avoided if possible, though, since the saving in cost is short term only, as some tapes are very abrasive and will wear out the heads of the machine very fast. Since the gap in the head is very small, even slight wear has a considerable effect, particularly causing loss of high frequencies. The only cure for worn heads is to replace them, and this can be rather expensive, particularly on high-quality machines. It would also be rather a waste to have a large collection of tape recordings made on cheap tapes which could not be played on a new machine at a later date, for fear of damaging it.

Another disadvantage of cheap tapes is that they suffer from what is called "drop-out". This is the effect of an uneven coating of magnetic oxide on the tape, which causes those parts of the tape poorly coated to fail to accept a recording properly. The result of this on playback is short periods of silence in the programme material, which can be very irritating.

It was stated earlier in this chapter that tapes would accept only a certain amount of magnetism before saturation and distortion, and that this set an upper limit on the level of the signal to be recorded. A few years ago a new type of magnetic coating, called chromium dioxide, was developed which will accept higher signal levels. This kind of tape gives a better signal-to-noise ratio than the normal ferric oxide tapes, since the audio signal can be recorded at a higher level compared to the noise generated by the tape. It does, however, require higher erase voltages and different recording conditions in the tape machine, and should not be used unless the machine is designed for it. There will be a switch on the machine if this option is available, usually marked CrO_2, in the position for this type of tape. Without the proper conditions, the result of using chromium dioxide tape will be worse than using normal tape, even though it is more expensive.

New types of tape are continually being developed to work better on non-CrO_2 machines. These usually involve finer processing and mixtures of magnetic oxides; they are more expensive, but the improvements are noticeable in lower background noise and in less susceptibility to overloading.

DOLBY SYSTEM

Another way of improving the signal-to-noise ratio is using the Dolby system, named after Ray Dolby, its inventor. In this system the audio signal is broken up into different ranges by electronic circuitry, and processed before recording. During playback the reverse process is applied, and any noise generated during the record/playback operations is reduced in this reverse process. The system is very effective, and a more sophisticated form of it is used by all large record companies now during the making of records – this accounts in part for the improvement in the quality of records in recent years. Tapes recorded using the Dolby system should be played back on machines with Dolby facilities or the results can be strange. Once again a listening test is the best way for the user to decide whether the extra cost of a tape machine with the Dolby system is justified.

Other systems of noise reduction are available and some are equally as good as (or better than) the Dolby one, if not so well known. They all operate in a similar manner to that described above. If another system is offered, listen to it before making a decision.

Summary

1. Check whether you want cassette, cartridge or "reel to reel". Cassettes are most convenient and are improving

The Tape Recorder

in quality. Cartridges are not hi-fi but give long playing times. "Reel to reel" gives flexibility for editing tapes, etc., but is less easy to store; possibly this medium is more suitable for the enthusiast (whether established or in embryo).

2. Check the track format and speeds are suitable for your needs. The more tracks the lower the fidelity, but less money is spent on tapes, similarly for low speeds. With cassettes the speed is constant and the choice is simply between mono and stereo.
3. Make sure that you can record on the machine you buy if you wish to do so. Some cassette machines are players only and are used with pre-recorded tapes.
4. If high fidelity is maintained, tapes are approximately as expensive as records: don't think a tape recorder will save you money.
5. Work out if you need a noise reduction system.
6. If you are going to do your own recordings, read further on the subject.

8 | SETTING UP A HI-FI SYSTEM

SURFACES

Having chosen and bought a hi-fi system and brought it home, there comes the problem of where is best to use it. All parts of the system should be placed on solid surfaces, whether these are shelves or tables. In this context a solid surface means one which is not vibrated during the normal course of using the room it is in; so a wobbly table will not do, nor will a firm one if this stands on a springy floor.

It is best to fix shelves to a brick wall and mount the parts of the system on these. Make sure that air can circulate round the power amplifier to prevent it overheating. Loudspeakers should not stand on the same shelf as a record player, as at high volumes they may vibrate the deck and cause an unpleasant howling noise, while at low volumes they will almost certainly cause resonances. There is no real reason why loudspeakers should even be on the same side of the room as the rest of the equipment.

WIRING

All the parts of the system should be wired together with screened cable (coaxial cable, or coax), except the loudspeaker lines. The use of coax prevents stray electrical signals such as mains hum which radiates from all mains wiring, and strong radio signals from nearby

Setting up a Hi-Fi System

transmitters (BBC, police, taxis, etc.) from getting into the system with the audio ones.

Loudspeaker lines carry powerful signals so that stray signals, which have a much lower power, are unnoticeable when mixed in with them. Since the power delivered to the loudspeakers is great, they should be wired with the heaviest quality of two-core mains cable that is convenient. This prevents signal being dissipated in the resistance of the wire, particularly on long runs of cable. (The resistance of wire gets less as the conductor gets thicker.)

Loudspeakers should never be wired with three-core cable using one of the cores as the common return for both speakers. This would allow the signal flowing through one loudspeaker to have a choice of return paths back to the amplifier, *i.e.* down the common wire or through the other loudspeaker and along its feed wire. The result is that some of the signal from one loudspeaker goes through the other and vice versa, giving bad crosstalk and therefore poor stereo. Errors of this kind in wiring may also damage the power amplifier.

All signal cables should be soldered correctly into plugs which fit the connections to the various parts of the system. Twisted wires do not make adequate connections. They may work for a time, but they fall apart easily and can also act as radio receivers, especially when they have been left for a while and have become slightly oxidised by the air. If loudspeaker connections are made with screw connectors, these should be done up firmly and care should be taken that whiskers of wire do not bridge the gaps between different contacts, as these will stop the system working and may cause damage to the electronics.

Various types of standard connectors are used, principally d.i.n. connectors, phono plugs and sockets, and coaxial plugs and sockets. Descriptions of these and

Setting up a Hi-Fi System

standard connections are in the Glossary at the end of the book.

Most Continental equipment is built to a specification for safety laid down by the International Electrotechnical Commission, known as IEC 65. This insists on an insulation standard known as Class II which in effect is double insulation and equipment built to this standard requires no earth connection. Two wires only will be found in the power cord and a label should be attached to this cord indicating the correct connection method.

NB: Consult a qualified engineer if in doubt about the safety aspects of your equipment.

EARTH LOOPS

Where not built to the Continental specification mentioned above, equipment must be adequately earthed or it will be dangerous – however, earth loops should be avoided. An earth loop is a continuous circuit in the earth wiring which enables a different path to be taken from the earth pin of the mains plug to a piece of equipment and back again from that equipment. The earth pin of a second plug may be part of this continuous circuit, since this will be connected to the first plug in the mains wiring.

The simplest way to avoid earth loops is to earth only the amplifier to the mains and to earth other pieces of equipment to the amplifier. This is frequently done using the screens of the audio leads connecting the pieces of equipment together. Permissible and incorrect connections are shown in Fig. 29.

For safety reasons, disconnect equipment from the mains by unplugging it before attempting any change in earthing systems. This means disconnecting all parts of the systems from the mains if they are joined by any type of wire at all.

Setting up a Hi-Fi System

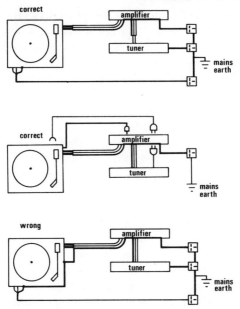

FIG. 29. Permissible and incorrect methods of avoiding earth loops

LOUDSPEAKER PHASING

The correct phasing of loudspeakers is important; that is they must both move in the same direction for the same signal. When correctly phased, a mono signal fed to a stereo system should sound fixed in a position equidistant between the loudspeakers. If the speakers are moving in opposite directions for the same signal, *i.e.* they are in antiphase, the signal will sound diffused. A pure tone fed to antiphase speakers will be interpreted by the listener as coming from some position, not necessarily equidistant between the speakers, but dependent on the frequency of the tone. As normal sound is a mixture of a large number of different pure tones, then the sound will appear to

Setting up a Hi-Fi System

come from a diffuse area. On antiphase speakers, therefore, an orchestra becomes completely nebulous, even appearing to come from points not between the loudspeakers, Even a single instrument will seem to be playing at several different positions.

To prevent this, loudspeaker wires should be identified in some way, usually by colour, so that it can be easily seen which wire of a loudspeaker pair connects which output of the amplifier to which terminal of the loudspeaker. One of the loudspeaker terminals will be marked, usually with a + sign or a red dot. If this terminal for both speakers is connected to, say, the brown wire of a brown and blue two-core mains cable, and this wire is also connected to the live part of the loudspeaker output connector on the amplifier, then phasing should be correct. Colours and marking systems will change from system to system but the principle remains the same.

To check phasing if it seems that it may be incorrect, play music with a lot of treble content on the system with the amplifier switched to mono and listen to hear if the music appears to be coming from a central position between the loudspeakers. Turn the music and amplifier off, reverse the wire connections on *one loudspeaker only*, turn the music and amplifier on again and repeat the test. Leave the speakers wired in the way that gave the best mono image.

POSITIONING IN THE ROOM

Loudspeakers should be positioned in a room in such a way that they are about 6 to 8 feet apart and have no furniture in front of them, particularly sound-absorbent things such as armchairs, sofas, etc. They should also be placed in symmetrical surroundings for an equal response from both, so that, if one is in a corner, the other should also be; if one is next to a curtain, the other should be, too, etc. If the room is rectangular, the speakers should

Setting up a Hi-Fi System

be against a short wall to take full advantage of the better bass response from using the length of a long room. Ideally the room should contain no right angles and no two walls should be the same size in order to avoid resonances. In modern houses of course this is often not possible, but resonances can be absorbed by heavy curtains, soft furniture and carpets in the room. If electrostatic bass units are being used, these should stand well clear of the wall behind them. It is quite permissible to conceal wiring other than mains under carpets, etc., but cleaning the carpet or other pressures will eventually show up the cable runs, so the wires should be moved every now and then.

In the final analysis the best arrangement of the parts of a hi-fi system is a matter of personal preference, so use this chapter as a guide on how to start and the considerations to be borne in mind, and experiment from there. Unless a room is specifically set aside as a music room it is unlikely that the hi-fi will be the only factor determining its arrangement, so the target is to achieve the best sound reproduction from an arrangement which is also convenient.

9 | MAINTAINING A HI-FI SYSTEM

GENERAL

A good hi-fi system, including the record collection, will be one of your more expensive possessions. It is delicate and requires careful maintenance to preserve it – it is surprising how badly some are kept.

The first rule of maintenance is to follow the manufacturers' instructions. To this end, read all instructions carefully when you first buy the equipment and then keep all leaflets and handbooks in a safe place so that they do not get lost and can be easily referred to. It is often convenient to put them with the record collection, for instance.

It is good practice to clean and oil (but only where the manufacturer tells you to) all the equipment about once every three months using the correct materials for this job.

Dust is one of the prime enemies of records, record players and tape recorders. Also, candles and coal fires in a room will quickly coat records and equipment with grime if they are not covered up. So keep records in their sleeves and put the lids on record players and tape recorders when they are not in use.

RECORD PLAYERS

Periodic checks should be made on the stylus assembly of a record player, since even a diamond stylus will wear

Maintaining a Hi-Fi System

out after a while. Many shops have a microscope for checking stylus wear. There are also proprietary fluids for cleaning dust and grease off the tip of the stylus assembly when necessary. Again, follow the manufacturers' instructions when cleaning. If you wish to check that signal is getting from the cartridge to the amplifier, the stylus may be rubbed gently with a single hair and a scratching noise should be heard from the loudspeakers. Do not rub the stylus with your finger, as this will put grease on it which will collect dust and damage records.

RECORDS

Records should be handled only by their edges or the centre label. Do *not* touch the grooves with your fingers as grease will get into them and collect dust. Even before collecting dust, finger marks will be heard on any reasonable equipment and they are difficult to get rid of since even careful rubbing with a record-cleaning pad will often just push the grease to the bottom of the groove, making it harder to remove.

If a record has become unbearably dirty it may be cleaned in a very weak solution of washing-up liquid using a soft lint-free cloth rubbed along the direction of the grooves, and rinsing with distilled water. A good method of drying records after this is to thread them on a string separated from each other by cotton reels, and then hang them up so that the string hangs vertically. Do be careful, however, that the records do not knock into each other during the threading, and also make sure that the string is strong and firmly suspended since it will be rather heavy with a large number of records hanging on it.

Do not smoke near records since cigarette smoke is greasy and ash falling on the record usually has enough heat left in it to damage the record surface, leaving audible clicks and bumps. Also avoid sneezing or coughing over records as the resulting deposits often cause

Maintaining a Hi-Fi System

groove jumping and are very difficult to clean off. If a record gets scratched there is no cure.

Records should be kept in their inner sleeve inside their dust cover, with the entrance of the inner sleeve at right angles to the entrance of the dust cover so that they cannot roll out by accident. They should be stacked upright in groups of less than twenty, and should be well supported so that they do not lean at an angle to the vertical. This prevents them slowly bending either under their own weight or under that of those leaning against them. Keep records away from heat and direct sunshine, as these will warp them and it is a slow and chancy business to reverse the warping process. Never leave records piled on top of one another, as any dust that has got inside an inner sleeve may be forced on to the surface of a record and the weight of just one other record on top of it will leave an audible mark after only a few weeks. Dust particles are frequently larger than the audio signal on the groove so they can have quite a large effect, one particle often causing clicks for three or four revolutions of the record.

Records should be cleaned each time they are played by wiping with a piece of dry, clean velvet, or one of the appliances sold for this job. Again, follow the instructions. The use of one of the devices which fit on to a record deck to clean records while they are being played helps to keep them in good condition. Do, however, keep the device itself clean or it will have no good effect. Periodically records should be given anti-static treatment to prevent build-up of static electricity which attracts dust.

Records should be kept on a shelf rather than on the floor as they will be less likely to collect dust. Also it is a good idea to keep them in some kind of order right from the start with a small collection, as this prevents damage when searching for a record and saves a lot of time when the collection gets large.

Maintaining a Hi-Fi System

TAPE RECORDERS

Tape recorders should be kept clean in much the same way as record players. Buy a kit of tools to keep the heads clean, but if this is not available they may be cleaned with cotton wool wound on a matchstick and dipped in methylated spirits. Do not bring metal near tape heads as there is a danger of scratching them or magnetising them, giving poor reproduction. Every now and then tape heads should be demagnetised using an electric tool sold for this job. Again, follow the instructions. Alternatively the service department of a hi-fi shop should be able to do this for you. Failure to demagnetise occasionally results in distortion, high-frequency losses or noisy playback.

TAPES

Tapes themselves should be kept in their boxes away from heat and magnetic fields, *e.g.* loudspeakers, powerful electrical equipment, etc. Apart from this they are reasonably robust, though care should be taken not to crease the tape itself or get finger grease on it. If you do have a tape-head demagnetiser, do not leave it switched on near tapes as it demagnetises them, too, thus erasing the signal. Obviously care should be taken with cartridges and cassettes since these are delicate mechanisms and should be kept free from dust.

10 | THE NATURE OF SOUND

CAUSE AND PROPAGATION

Sounds are small vibrations in the air. The air does not move any distance as it would in the wind, but merely oscillates with respect to its surroundings. A sound is caused either by something vibrating which is in contact with the air, as when a surface is banged or a string plucked, or by an air stream being unstable and wobbling around as it flows, as with wind instruments or the noise of wind in a moving car. With wind instruments the instability of the airflow is carefully controlled to produce a musical note.

As each small part of the air vibrates, it affects the part next to it, which vibrates in sympathy and affects that next to it and so on. If the sound source stops, however, the oscillations die down very quickly.

It takes a finite length of time for the sound to travel – it takes a little less than one-thousandth of a second to travel one foot – so if a listener is one hundred feet away from the source, as perhaps in a concert hall, he will hear it with a delay of just under a tenth of a second. When the sound reaches the ear, it vibrates the ear drum. This converts it to a signal in the nervous system, which is what the brain "hears".

ENERGY AND HEARING SENSITIVITY

A single short sound from a given source can contain

The Nature of Sound

only as much energy as it was originally given. As the sound travels away from the source its energy can be considered to be spread over the surface of a sphere which is getting larger and larger at the speed of sound, so the amount of energy per square inch of the surface gets less and less. Since the area of the surface of the sphere is proportional to the square of the radius, the sound is spread out four times as thinly each time the listener doubles his distance from the source. This is an example of the inverse square law. The human ear compensates for this by being more sensitive to quiet sounds than loud ones, though it operates on average levels so that a quiet sound would get lost in the middle of a loud crash when the average level was high.

In order to describe this change in sensitivity without involving enormous numbers at high volumes to express slight effects as far as the listener is concerned, the decibel scale has been worked out. Originally the unit was one bel, but this was found to be too large for convenience, so decibels are used where 10 decibels (10 dB) are equal to 1 bel (1 B). A decibel is a ratio, that is it is always a description of something with respect to something else. Thus 10 decibels is a ratio of 10:1; minus 10 decibels is a ratio of 1:10 (a positive sign means that it is bigger, a negative sign that it is smaller). Twenty decibels is two ratios of 10:1, *i.e.* 100:1 (the law works on logarithms, *i.e.* $x\,dB = 10\log_{10}R$, where R is the ratio of the two numbers being considered; in this case $R = 100$, $\log_{10}100 = 2$, $x = 20$).

Three decibels is a ratio of 2:1 and is used in describing frequency properties of amplifiers. The 3 dB point is the frequency at which the fall-off in frequency response in the treble or the bass has reduced the output power to one half of its middle frequency value, and is a convenient standard.

A standard level of sound is used in measurement, and

The Nature of Sound

a decibel meter gives the ratio, expressed in decibels, of the volume of sound being measured to this standard. So a sound pressure level of 23 dB means that there is 200 times as much power being received by the meter than the standard reference level. The standard is approximately equal to the quietest noise a human ear can perceive.

Over most of the human hearing range from full sensitivity to least, a change of a given number of decibels will sound the same. So the difference in volume between 0.10 watt and 1 watt will sound the same as the difference between 3 watts and 30 watts, it being 10 dB in both cases, though the latter started off louder. There is a bottom end to the sensitivity of hearing past which a sound is too quiet to hear, and a top end at which a sound is so loud as to cause physical pain. The figure normally accepted for the latter is $+120$ dB; the former depends on the hearing of the individual.

As the sound's volume gets lower and the ear increases sensitivity to hear it, so the listener's frequency range changes and at low volumes the treble and bass tend to get lost. It is a good idea when listening to hi-fi at low volumes to increase the settings of the treble and bass controls slightly to compensate for this. Alternatively, if it is fitted, use the loudness control, which has the same effect.

The effect of distance on sound is not only that it gets quieter as one moves away from the source but also that the bass is progressively lost. The transmission of sounds through the air is less efficient in the bass and it tends to get damped out before it travels very far.

WAVEFORMS

Sound was described as slight oscillations of the air. These movements are accompanied by changes in pressure which follow exactly the same patterns as the movements.

The Nature of Sound

A graph of pressure against time can be drawn, and this will show in visual form what the sound is like. This is called the "waveform". The purest wave is a sine wave, which is shown in Fig. 30.

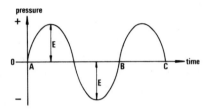

FIG. 30. Sine wave

Note that it swings equally above and below the zero pressure line or else there would be a residual pressure left which would cause permanent movement of the air. One cycle of this wave is one complete shape before the curve exactly repeats itself, *i.e.* the curve between A and B; the curve between B and C is a half cycle.

The pitch of the sound is determined by the frequency of the waveform. If the time scale is such that the distance from A to B equals one millisecond (one-thousandth of a second), then the frequency is 1000 cycles per second. This is the same as 1000 hertz (1000 Hz or 1 kHz). The distance E shown on the graph is called the "amplitude". All sounds consist of a combination of sine waves of different frequencies and amplitudes.

Even the sound of a symphony orchestra could in theory be broken down into a very complicated mathematical expression which described it exactly in terms of lots of sine waves of different frequencies and amplitudes, but the expression would change all the time the orchestra was playing. However, there is a use for all this. The human ear has a limited frequency response, say 20 Hz to 20 kHz, though most people are more restricted than this. If a hi-fi system responds equally to all sine waves in this frequency range, then it will be able to reproduce any waveform that the human ear can hear, no matter what the complexity. So sine waves are used to test equipment and can be used to define the quality of the equipment.

The word "phase" is used to describe the relative timing of two sine waves to each other. They must be of equal frequency for their phase to remain constant, since the phase describes how far through a cycle one of them is before the other starts a new cycle (*see* Fig. 31).

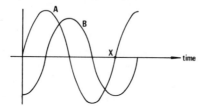

FIG. 31. Two sine waves with phase difference

Sine wave B is phase retarded (it happens later) on sine wave A. They have the same frequency as each other but their amplitudes are different. Amplitude differences do not affect phase relationships as it is the horizontal axis which defines phase. A cycle is divided into 360°, so since wave B starts a quarter of a cycle after wave A it is 90° retarded. (Note that this could be 270° advanced if the point X were viewed as the starting point for wave A – the expressions are mathematically identical if A and B are continuous waves.) Fig. 32 shows 180° phase angle,

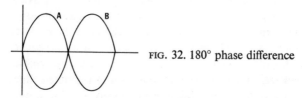

FIG. 32. 180° phase difference

The Nature of Sound

which is often called "antiphase". At any time the two curves are moving in opposite directions. "In phase" means that the two curves cross the zero line at the same places in the same directions.

The wavelength describes the distance travelled by a wave of a given frequency between the beginning of one cycle and the beginning of the next. The speed of sound is, quite accurately, 1100 feet per second, so, if the frequency being considered is 100 Hz, then the wavelength is 11 feet (one cycle is one-hundredth of a second in duration and the sound has travelled 11 feet in that time). This is reasonably important in choosing a loudspeaker for a room, since when the wavelength of the sound gets longer than the length of the room, the speaker/room combination becomes inefficient at producing sound; *i.e.* from the example above a speaker in a room 11 feet long would have a falling response below 100 Hz, no matter how good is the inherent bass response of the speaker, and frequently how high its price.

DISTORTION

Distortion is a feature of all sound-reproducing equipment. In theory, the signal recorded on a record or tape or transmitted by radio is as near an exact copy of the original sound as possible. The reproducing equipment should distort it as little as possible so that the only alterations to it are those controlled by the listener with the tone controls, volume controls, etc. In practice all equipment will change the waveforms very slightly, and these changes will always be in the form of harmonics of the original sine waves in the programme material. A harmonic is a signal at some multiple of the basic or fundamental frequency, so the harmonics of 100 Hz are 200, 300, 400 Hz, etc. The distinctive characteristics of musical instruments are due to harmonics of the notes they are playing. The different proportions of harmonics make, for instance, a clarinet different in sound to an oboe, though they may be playing the same note.

A hi-fi system should not introduce any extra characteristics to the music being played on it, as these were not intended on the original recording, and anyway usually sound pretty awful! The distortion of a system is measured by putting a pure sine wave at a given frequency into the input and then filtering the output to remove only that frequency. Other frequencies generated by the system itself get past the filter and are measured on a meter as a percentage of the unfiltered output. Hence distortion is expressed in per cent.

The distortion components at twice and four times the original frequency are related musically to the fundamental (they are one and two octaves above it), so do not sound so noticeable as that at three times the original frequency which bears no musical relationship to the fundamental but only a mathematical one. Higher harmonics tend to be of very low amplitude.

There is also a form of distortion called "crossover" distortion which is inherent in certain types of circuit. This puts a little kink in the waveform at the point it crosses through zero on the graphs in Figs. 30 and 31. It is surprisingly audible and very harsh as a sort of buzz.

Although to an untrained ear distortion as high as 1 per cent may not be noticeable other than as a feeling that all is not entirely well, as it becomes used to better- and better-quality equipment the tolerance level reduces to about 0.1 per cent as a limiting value. For this reason it is worth buying as good equipment as possible at the outset so that the sensitivity of your ear to distortion does not overtake the quality of your equipment.

STEREO

In mono equipment, a single loudspeaker plays all the music so that it comes from one place. Although all the

The Nature of Sound

notes are there, and there may be some feeling of depth and distance from the tonal quality of the sound, there is no real positional information.

In stereo the two loudspeakers generate different signals so that different things are heard by each of the listener's ears. The differences may be subtle but are still important.

When a sound comes from immediately in front of the listener, identical signals are heard by each ear. This is the mono condition of a stereo amplifier or signal from a mono amplifier fed to two loudspeakers. As a signal moves to one side, firstly there is a difference in volume received by each ear, and secondly the phases of the signals received by one ear are different from those received by the other. This is because the distance travelled to the ear away from the sound source is further than that to the ear nearer to it, and there are obstacles in the way as well which modify the sound (nose, hair, etc.). The brain is very adept at positioning sounds from this information – it has after all had a lifetime's practice.

With a stereo system, the signals from the two loudspeakers carefully re-create the differences in phase and volume and build up an image in sound that stretches between and behind the loudspeakers. If the speakers are wired antiphase, then this information is at variance with itself and no coherent image results.

ECHO

If a sound travels a distance longer than about 25 feet and bounces off a reflective surface, then a listener standing next to the source will be able to hear not only the original sound but also the reflected sound as a separate source. The further the sound travels before reflection, the longer and more noticeable the gap between the original and the returned signal. This is echo. If the distance is less so that the sound returns in less than about 40 milliseconds,

The Nature of Sound

the brain cannot separate the two, and the result is a slight blurring. However, if the original is in one ear and the return is in the other, it may still be possible to separate them spatially, even at such short time intervals.

Echo is often used on records, particularly on popular music, as a special effect, but it should not occur in a listening room as it will change the stereo image by adding different phase relationships to the sounds received by the listener's ears. To this end, a room containing a hi-fi system should have acoustically damped walls.

For certain frequencies, with undamped surroundings, the returning echos may always be in phase or in antiphase on the original signal, depending on the dimensions of the room. In the former case they will exaggerate it; in the latter case they will deplete it. This is known as "resonance" and "anti-resonance" and can add an unpleasant booming sound to music or give dead spots in the frequency response. Normal furnishings will tend to minimise these effects, but, if trouble is encountered, speakers should be repositioned to avoid them, particularly signals being bounced back and forth between parallel walls of a room. Heavy curtains will often help.

REVERBERATION

Reverberation is similar to echo in that signals are bouncing off walls, but the time taken to return is so short that the delay is not noticeable, and there are so many paths the sound can take that a continuous reflected signal is heard over some time. All rooms have this; without it they would sound dead.

The time that the sound takes to die to one-thousandth of its original volume is called the "reverberation time", and for domestic surroundings in rooms about 12 feet square it is in the order of one to one and a half seconds. The ideal amount of reverberation in a listening room is a matter of preference, but one should be aware that it matters and experiment if possible to find the best amount. In general, heavy furnishings and curtains in a room will shorten the reverberation time, but the dimensions of the room determine its tonal quality to a great extent.

11 | RECORDING

TAPE RECORDER LEVELS

The primary purpose of recording is to get a well-composed sound picture with as few defects as possible. In order to do this the tape machine must be in as good a condition as possible, and microphones, if the recording is live, must be chosen and used carefully.

Recordings made from records will usually infringe copyright regulations, but, where the records are of sound effects, this is generally not the case. If there is any doubt as to the legality of recording material further details may be obtained from the Performing Right Society, 29–33 Berners Street, London W1. A tape recording from a record should be made using a cable to connect the tape output of the amplifier to a suitable input on the tape machine. The input sensitivity, or recording level, control of the tape machine should be adjusted so that the loudest part of the record deflects the record level meters just past the "0" or "100 per cent" mark (*i.e.* into the red portion of the meter), but never so that it remains there continuously. If it does remain in the red for long periods, the signal on the tape will probably be audibly distorted. This is a guideline only; experiment will show what is best with specific machines – they vary greatly.

If the meters show too low a level is being recorded, then the background noise of the tape will interfere unnecessarily with the material being recorded. The recording level controls should not be adjusted once the recording has started if this is at all possible, but, if it is felt to be necessary, all changes should be made very slowly indeed or they will be apparent on the finished recording.

Some machines have automatic level controls which prevent very loud programme from overloading the tape. These work satisfactorily only if the input level fed to the machine matches the sensitivity of the input amplifier; if not, then very strange effects occur which often spoil a recording completely. Such devices should be used with care, though if correctly used they take a lot of the worry out of making recordings.

If the tape recorder is a three-head machine, then it will be possible to monitor the signal on the tape as well as that being fed to the tape. This gives an audible quality check on the recording during the recording process. This is invaluable since any errors can be stopped as they occur rather than at a later time which is often inconvenient.

The competent use of a tape machine to give good recordings is something that comes only with experience, and it is probably best to practise with records before trying anything ambitious with microphones, though passable live recordings are possible from the word go.

MICROPHONES

The whole subject of microphones and their use is very complex, so only the basics will be described here. If further reading material is required there are several books which cover the subject in greater detail. I would recommend *The Technique of the Sound Studio* by Alec Nisbett, published by Focal Press, which also gives a great deal of useful information on allied subjects.

There are two basic types of microphone: velocity and pressure microphones. Velocity microphones pick up the

Recording

oscillations of the air, and the sensing element is carried with these vibrations and converts the oscillations into an electrical signal. If the oscillations are at right angles to the direction of motion of the microphone element, they cannot move it, and so these microphones have a directional response. This means that they are more sensitive to sounds from some directions than from others. The simplest form of velocity microphone is a ribbon microphone, where a very light ribbon of metal is suspended between the poles of a magnet (*see* Fig. 33).

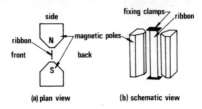

FIG. 33. Schematic of ribbon microphone

Sounds coming from the front or the back vibrate the ribbon, causing small electrical currents in it, which are the output. Sounds coming from the sides, however, cannot, not because the poles get in the way – the sound goes round them – but because they are coming from the wrong direction and catch the ribbon edge-on. A plot of the sensitivity of a ribbon microphone is shown in Fig. 34.

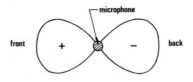

FIG. 34. Ribbon microphone polar response plot

The lines join all points of equal sensitivity round the microphone. This is a polar response plot. A sound from the back of the microphone produces an inverted signal to one from the front (*i.e.* antiphase), hence the + and − symbols. This type of response is called a figure-of-eight for obvious reasons.

A pressure microphone operates by the changes of air pressure due to the sound vibrating its sensing element. The sound can approach only one side of the element, the other being sealed off from it to provide a reference. The simplest form of pressure microphone is probably a moving-coil one built similarly to Fig. 35.

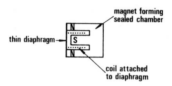

FIG. 35. Schematic of moving coil microphone

Sounds vibrating the diaphragm cause the coil to oscillate in the field of the magnet, generating small electrical currents which are the output. A sound coming from any direction will vibrate the diaphragm, so the microphone is called omni-directional. Its polar response plot is shown in Fig. 36. The entire plot has the same phasing.

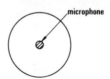

FIG. 36. Polar response of omni-directional microphone

If the ribbon microphone is used in the same place as the omni-directional one and their outputs are added together electrically, then the signals from sounds coming from the back of the ribbon will cancel with the signals from the same sounds from the omni-directional microphone, since they are of opposite phase, whereas those

Recording

from the front will reinforce. The resultant combined plot is shown in Fig. 37. This combination plot is called "cardioid", because of its heart shape, or "uni-directional", since sounds from the front are picked up but those from the back are not.

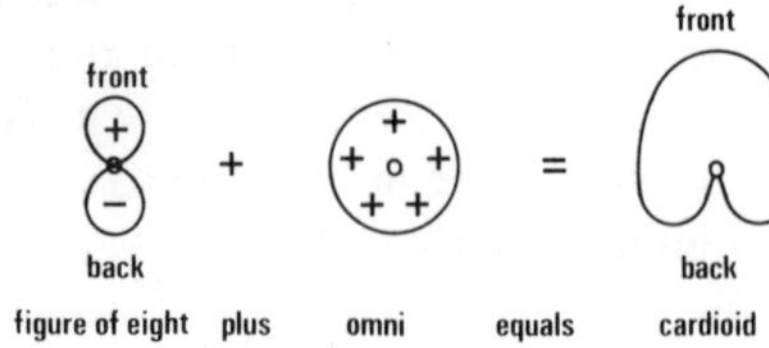

FIG. 37. Combining microphones for a cardioid response

The types of microphone described above are by no means the only ones available, though one of the three basic plots is always descriptive of the response. There are different plots available, but these are specialised versions of those already explained. Hypercardioid is like cardioid with less pick-up from the sides, but usually this means more from the back (*see* Fig. 38).

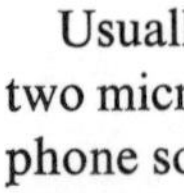

FIG. 38. One type of hypercardioid polar response

Usually a cardioid response is not made by combining two microphones, but by designing the body of the microphone so that the sound paths within it give the response using, say, a single moving coil element. These microphones usually have vents just behind the front and, if these are covered up, the microphone reverts to omni-directional operation.

Polar response plots are accurate only at one frequency, so a microphone should be supplied with a set of response plots for a range of frequencies. Many microphones are omni-directional at low frequencies, but this does not matter too much since the human ear is also, and stereo effects will not seem unrealistic because of it.

If two cardioid microphones are placed next to each other pointing at 90° to each other in a horizontal plane (Fig. 39), then sound A will be in the field of one micro-

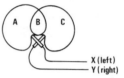

FIG. 39. A coincident stereo pair of cardioid microphones

phone, sound C in that of the other and sound B in both. Connect cable X to the left channel of a tape recorder and cable Y to the right. A recording of A, B and C will play back with the sounds coming from the same positions between the speakers as they occupied in front of the microphones. The set-up is called a coincident pair and is the basis of many stereo recordings.

If the same thing is done with ribbon microphones as in Fig. 40, then A, B and C will come out as before, but F will also be on the left channel, D also on the right channel and E in the centre. D and F have been reversed in position, and D, E and F have become mixed with C, B and A. Note that C and D are completely in the dead spots of

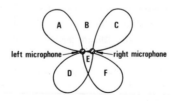

FIG. 40. A coincident stereo pair of "figures of eight" microphones

Recording

the left channel's microphone and will hardly be heard on the left channel of the recording.

Microphones can often be positioned with reference to their polar responses to mix sounds in interesting ways and isolate out unwanted sounds by placing them in the dead spots of the response. An example of when such isolation is used is when recording a guitarist where the microphone may need to be near the body of the guitar to pick up mellow tone, but the noise of the guitarist's fingers on the strings while changing cords is not required.

EDITING

A recorded tape will probably not be exactly as it is finally wanted; mistakes may have occurred, or it may have taken two or three attempts to get a good recording of the total programme. It is not good practice simply to erase a tape and record over the top of it if early attempts were unsuccessful. Erasure is never complete and a background, which is the remains of an early recording, may be heard on a later one.

The correct technique is to carry on with each "take" on unused tape, and then cut out the unwanted material using a razor blade and a splicing block, which acts as a vice to hold everything steady. The final step is to join the pieces of tape that contain the wanted material together using the splicing block and special splicing tape. Do not use any other type of adhesive tape as the glue used is not suitable. Above all read the instructions that come with the splicing block, and avoid using a razor blade that has become magnetised – this will put clicks on to the tape at each splice.

After practice, it should be possible even to splice one note into the middle of a tune from which it was missing, if the music is suitable, so that the splice is undetectable. In order to do this, the tape should be run past the playback head of the machine by hand so that the

Recording

exact places to cut are located accurately. The back of the tape is then marked with a yellow wax pencil immediately over the head gap at each required cut; the tape is removed from the machine to the splicing block and then cut. After splicing the required pieces together, taking care not to turn the tape round or over, play it through the machine to check quality. Do not despair if it comes out wrong the first few times – it can be a tricky operation. On particularly difficult musical splices it may be useful to mark the back of the tape with the positions of the beats of the music. If this is done, the timing through the splice can be checked with a ruler.

With tape recording, good results can be obtained early on as long as nothing too ambitious is tried. Learn to do the simple things without having to puzzle them out each time before trying for more subtle effects or complicated tricks.

GLOSSARY

a.c.: Alternating current. A current which cycles regularly from a positive to a negative value, usually as a sine wave. Also used of the voltage driving such a current. British mains are an a.c. electricity supply.

AFC: Automatic frequency control. A circuit to keep a radio tuner on station despite any drifting in the circuitry with temperature or time.

AGC: Automatic gain control. (*a*) A circuit to maintain a constant signal level in a tuner or radio, thus giving optimum reception of signals at different strengths (*see* Chapter 4). (*b*) Also a circuit to adjust automatically the level of a signal recorded on a tape recorder for optimum results without adjustment. The latter is sometimes called "automatic volume control" (AVC) or "automatic level control" (ALC) (page 107).

AM: Amplitude modulation (page 38).

Acoustic: (*a*) Having to do with sound; principally used of buildings. (*b*) The overall sound properties of something.

Aerial: A piece of wire or more complicated apparatus for receiving radio signals and feeding them to a radio tuner (page 44).

Ambiophony: A type of four-speaker system which produces sound from all round the listener (page 15).

Glossary

Amp: (*a*) The basic unit of electrical current: *see* Ohm's Law. (*b*) Abbreviation of "amplifier".

Amplifier: An electrical circuit in which a signal is modified to make it more suitable to drive following equipment (*see* Chapter 5).

Amplitude: The overall size of an electrical signal (page 99).

Audio: To do with sound; principally used of electronic equipment.

Automatic frequency control: *See* AFC.

Automatic gain control: *See* AGC.

Balance: (*a*) The relative volumes between the left and right channels of a stereo system to give a realistic stereo image. (*b*) A description of the relative volumes and tonal qualities of different instruments or voices in recorded or broadcast material.

Bandwidth: The difference between the lowest and highest frequencies a circuit will accept (page 13).

Bass: The lower end of the audio spectrum.

Bass reflex: A type of loudspeaker enclosure (page 63).

Belt drive: A method of driving a record player turntable where the motor is coupled to it through an elastic belt (page 33).

Bias: (*a*) In a record player this is a mechanical force on the arm which has to be offset (page 31). (*b*) Also a signal outside the audible range used in tape recorders to overcome inherent distortion and poor bandwidth in the recording process.

Bright: Used principally of loudspeakers to describe increased response in the upper-mid audio band.

CrO_2: Chromium dioxide. A magnetic coating used on recording tapes, giving higher signal-to-noise ratio than ferrous oxide (page 83).

Cabinet: The enclosure in which a loudspeaker is mounted.

Glossary

Capstan: The drive shaft on a tape recorder that draws the tape across the heads and regulates its speed (page 79).

Cartridge: (*a*) The actual transducer on the end of a record player's pick-up arm that converts the mechanical signal to an electronic one (page 23). (*b*) A method of storing and carrying tape in a small box with access for playback on a cartridge machine; this is not a cassette (page 71).

Cassette: A method of storing and carrying tape in a small box with access for record and playback on a cassette machine; this is not a cartridge (page 71).

Ceramic cartridge: One type of record player cartridge (page 23).

Channel: One of the two audio signal paths in a stereo audio system.

Circuit: (*a*) Any electronic path. (*b*) Also used of combinations of components assembled to perform a particular function.

Coax: *See* Coaxial cable.

Coaxial cable: A cable consisting of two conductors, one as a wire and one as a tube, both sharing the same axis. The tube is usually used as an electrical screen for the wire within it.

Coaxial connector: A specific type of plug and socket designed to maintain the geometry and screening properties of coaxial cable. Often used on radio and television aerials. *See* Fig. 41 for wiring details.

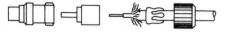

FIG. 41. Wiring of coaxial connector

Colour: Used to describe the sound of any specific loudspeaker as opposed to a theoretically perfect one.

Compatibility: (*a*) The ability of one particular system of signal processing to be used with apparatus designed for another system (page 15). (*b*) The ability of different items of electronic equipment to be used with each other. Rules-of-thumb are as follows:

1. For low signal levels (*i.e.* anything not driving a loudspeaker) voltage levels should be similar and impedances of inputs should be the same as or higher than those specified for the output connected to them.
2. A loudspeaker should be able to handle the output power of the amplifier driving it. It should have the correct (or higher) impedance for the amplifier.

Compliance: A measure of how easily a record player stylus will follow a signal on a record (page 26).

Crossover: An electrical circuit used in loudspeakers to divide the signal into the correct frequency bands for the individual loudspeaker drive units (page 59).

Crossover distortion: A type of very audible distortion in a power amplifier which happens as the signal swings from positive to negative (page 102).

Crosstalk: Signal from one channel getting through to the other in a stereo system. This is undesirable since it spoils the stereo image. Look for large negative decibel figures in a specification.

Current: The flow of electricity. The quantity flowing is measured in amps. *See* Ohm's Law.

Cutting lathe: The apparatus on which a record is cut in a recording studio.

Cycle: One complete oscillation of a waveform (page 99).

d.c.: Direct current. A current which maintains a constant value and direction. Also used of the voltage driving such a current. A battery supplies d.c.

d.i.n.: Initials of the German standards institution, responsible for the definition of many of the standards used in hi-fi equipment, and also methods of measurement of performance. The initials are used as the name of a range of connectors; the wiring of the most common five-pin one is given in Fig. 42.

Glossary

The three-pin version – used for mono – simply omits pins 4 and 5.

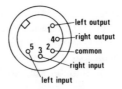

FIG. 42. Five-pin d.i.n. connector showing standard connections, swop inputs and outputs for tape machines

Deck: (*a*) The actual mechanical apparatus of a record player or tape recorder excluding its box and amplifiers. (*b*) Less rigorously, a separate record player or tape recorder without the parts normally contained in the integrated amplifier but with all parts necessary for its own function, including its box.

Decibels: A method of describing the ratio of one level of power to another. Derived from a "bel", the decibel is one-tenth part of it. A decibel figure is always a ratio. Sound pressure decibels are related to the quietest sound the human ear can hear: 0 dB means the power content is the same as the reference. Negative decibels relate to fractions, positive to multiples. The mathematical formula for decibels is:

$$N = 10 \log_{10} \frac{A}{B}$$

where N is the number of decibels;

A/B is the ratio being described.

Decode: The action of extracting an original audio signal from a different electrical or mechanical form in which it may have been stored or transmitted. More specifically used when the character of the signal has been altered, for instance to carry two or more signals on one signal path simultaneously.

De-emphasis: The action of circuits, usually in a radio receiver, to remove exaggeration put there to overcome system deficiencies in a transmitted signal. The exaggerating process is called "pre-emphasis" (page 42).

Detector: The circuit which extracts the audio signal from the processed radio frequency signal in an AM radio set.

Glossary

Dipole: A simple form of radio aerial consisting of two wires pointing in opposite directions (page 45).

Direct drive: A method of driving the turntable of a record player in which the platter is mounted directly to the drive shaft of the motor. This is a recent development commercially since manufacturing motors that rotate accurately at the slow speed required poses difficult engineering problems only lately solved.

Discriminator: The circuit which extracts the audio signal from the processed radio frequency signal in an FM tuner.

Distortion: Any alterations to a signal other than in amplitude or frequency response introduced by equipment. These alterations normally consist of the addition of harmonics of the original signal not intended to be there and frequently not musically related to it. They may also be caused by one part of the signal interfering with another through inaccuracies in the equipment (this is "intermodulation distortion)". Transient intermodulation distortion is caused by one part of the equipment becoming incapacitated for a short while by part of the signal and therefore being incapable of reproducing any of the rest of it until the trouble clears away. The times involved in TID are of the order of one-hundredth of a second (page 84).

Dolby: An electronic method of overcoming the inherent noise problem in recording or transmitting audio signals. Named after Ray Dolby, its inventor (page 84).

Double play tape: Recording tape made half the standard thickness enabling twice as much to be stored on a given size of reel. Naturally it is more fragile than standard play and suffers from print-through, but storage is more convenient.

Drift: A failing of simpler radio tuners and receivers, when they gradually drift off tune with time and temperature and have to be adjusted. Automatic frequency control obviates the need for hand adjustment when this happens by continuously making the adjustment electronically.

Driver: A term for the electromechanical element in a loudspeaker.

Glossary

Dropout: Gaps in a recording on tape due to imperfection in the tape or tape recorder mechanism.

Echo: Fast repetition of an audio signal either once or several times, but with an audible gap between each repetition.

Editing: The cutting, splicing and rearrangement of recording tapes.

Effective mass: Principally the mass operating on the tip of the stylus of a record player, including the effect of the balancing, and any decoupling of heavy parts of the pick-up arm. Its value should be as low as possible for good tracking of the record, but read Chapter 3. The same concept may be applied to any mechanical system.

Efficiency: Usually of loudspeakers. The acoustical power output compared to the electrical power input expressed as a percentage. For transmission line speakers it is approximately 2 per cent; infinite baffle and reflex 5 to 8 per cent; horn 10 to 30 per cent. There is usually a trade-off of efficiency against fidelity. As a point of interest, one acoustical watt is *very* loud.

Electro-magnetic: Describing the interaction of electricity and magnetism. When electrical oscillations happen in a wire, these are accompanied by magnetic oscillations round the wire, and the two together spread outwards from the wire and can be detected. This is the basis of radio. The frequencies usable can be very high indeed and eventually those of light are reached. Light is electro-magnetic radiation.

Electrons: The smallest component parts of an atom. They have a negative charge and their movement is an electric current. Static electricity is a collection or absence of them in one place, usually on the surface of an insulator such as a record. In order to maintain a stable state, static electricity will tend to attract particles charged oppositely to itself and so neutralise itself; hence records collect dust if the static electricity is not removed from them. Friction causes static electricity build-up.

Glossary

Electrostatic speakers: Loudspeakers in which a thin insulating sheet is charged with static electricity and placed in a voltage field which oscillates with an audio signal. The effect is to attract and repel the sheet, making it give off sound waves (page 65).

Elliptical stylus: A particular shape of stylus which approximates as closely as possible to that of the original equipment used in making a record to give highest fidelity (page 26).

Enclosure: The box, and its detail design, in which a loudspeaker is mounted.

Encode: To change the nature of a signal in order to transmit, record or store it in a convenient form for distant or later replay through decoding equipment.

Equalisation: (*a*) Reinstating the original frequency response of a signal after it has necessarily been changed during a recording process. (*b*) Also a professional term for complex tone control systems (page 49).

Erase head: The part of a tape recorder that erases any previously recorded signals on a tape to ensure a clean background for a new recording (page 72).

FM: Frequency modulation. A method of encoding an audio signal on to a radio wave in order to transmit it. It is this type of radio transmission that is used for hi-fi broadcasts (page 39).

Feedback: A technique of feeding a fraction of the output of a circuit back to its input in order to check that the output is a faithful reproduction of the input. This is negative feedback, since errors tend to cancel in the process.

Positive feedback causes errors to exaggerate themselves and leads to instability. An example of this is a microphone held too close to a loudspeaker that is being fed from it, thus causing oscillations.

Ferrite: A magnetic material with very useful properties, sometimes used for tape recorder heads.

Ferrous oxide: The brown coating of magnetic material on a

Glossary

recording tape, recently being superseded in part by chromium dioxide (page 83).

Field: Any volume in which the magnetic or electrical properties are being considered; the interaction of electrical and magnetic forces within a volume.

Filter: An electronic circuit which allows only certain signal frequencies to pass through it, or alternatively prevents certain frequencies from passing. No circuit will act suddenly at one frequency and only allow, say, frequencies higher than that to pass, but not those that are lower. The rate at which it attenuates signal with change in frequency is called its "slope", and different slopes are required for different jobs (page 52).

Flangeing: A form of very fast echo, in which the gaps between each return of the signal are not audible. This is used in pop music.

Flat: (*a*) Of music, a note at slightly too low a frequency. (*b*) Of frequency response, that condition when all frequencies are reproduced with equal amplification.

Flutter: Fast variations in the speed of play of a record player or tape machine giving a fluttering effect (page 33).

Frequency: The number of times an oscillating signal repeats itself in one second (page 99).

Frequency response: (*a*) The range from the lowest to the highest frequency a system is capable of reproducing with equal amplification. This is usually measured between the points at which the power has dropped to half of its midrange value, since this is only just noticeable to the human ear. (*b*) Also used to describe the variation of amplification with frequency when it is intended not to be equal to all frequencies.

Gain: The relationship between the input signal and the output signal of an amplifier or system, measured in power, volts or amps.

Gramophone: Old term for record player, usually applied to mechanical ones.

Glossary

Harmonics: Whole multiples of a basic frequency; *i.e.* if F is the fundamental, 2F is the second harmonic, 3F is the third, etc. 2F, 4F and 8F are musically related to the fundamental; 3F and 5F are not. Since distortion is derived from a basic signal it tends to contain harmonics of that signal and so percentages are sometimes quoted in terms of the harmonics (page 101).

Head: The part of a tape recorder (or less usually a record player) that converts the recorded signal into an electronic one, or vice versa (page 72).

Headphones: Small loudspeakers mounted on a band to be worn on the head, positioning the loudspeakers next to the ears for private listening (page 68).

Headshell: The part of a record player's pick-up arm that carries the cartridge and stylus (page 30).

Hertz: The basic unit of frequency: 1 Hz equals one cycle per second; 1000 Hz = 1 kHz = 1000 cycles per second.

Heterodyne: A method of beating a frequency against another electronically to produce a third. This is used as the basis of very tight filters in radio tuners (page 38).

High fidelity, Hi-fi: The art of reproducing an original sound faithfully despite any transmitting or recording processes used in the interim.

Horn speaker: A very efficient, if bulky, type of loudspeaker (page 64).

Hum: Spurious signal injected into, or picked up by, electronic equipment from the local mains supply.

IC: Integrated circuit. A type of circuit in which large parts are made in one overall package, so reducing their size and lowering the number of components used. This increases reliability while reducing design flexibility, so that a trade-off is made between simplicity and high quality.

IF: Intermediate frequency. A frequency generated between the radio frequency being received and the audio frequency

Glossary

output of a radio or tuner. It is at this frequency that the filtering and amplification circuits operate (page 38).

Impedance: A general term for the resistance a circuit presents to an electrical signal, taking into account any phase-changing properties it may have (page 16).

Infinite baffle: Literally a board going off to infinity in all directions with a loudspeaker mounted in it. Used to describe a type of loudspeaker enclosure which is a closed box so that there is no sound path from the back of the loudspeaker to the front (page 62).

Interference: Any unwanted signal.

Jack, Jack plug: A type of connector consisting of a quarter inch segmented shaft which fits a socket consisting of a quarter inch hole with sprung contacts inside which make on to the various segments of the plug. For standard connections to the plug, *see* Fig. 43.

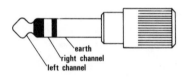

FIG. 43. Standard connections to jack plug

LW: Long wave. The lowest frequency band in general use for AM radio transmissions (page 38).

Limiter: A circuit which acts as a very fast electronic volume control and prevents a signal from getting large enough to distort in following circuits.

Linear, Non-linear: Descriptive of frequency response, meaning having equal amplification to all frequencies, or not.

Long play tape: Recording tape that is two-thirds the thickness of standard play tape, so preserving some of the robustness of the latter but improving on playing time for a given size of reel.

Glossary

Loudness: This is not volume, but an expression of the fact that at low volumes the human ear is less sensitive to high and low frequencies. A loudness control compensates for this (page 53).

Loudspeaker: The apparatus that turns electrical signals back into audible ones (*see* Chapter 6).

Loudspeaker phasing: The relative motions of the cones of loudspeakers when driven with the same signal. These should be identical (page 89).

MW: Medium wave. The middle frequency band in general use for AM radio transmissions (page 38).

Magnetic cartridge: A record player cartridge using the fact that a magnetic field moving across a conductor causes a current to flow to provide an electronic signal from the mechanical one stored on a record (page 24).

Matching: Making sure that all the electrical properties of various pieces of equipment are compatible. Also that the quality of each part of a hi-fi system is consistent with the equipment with which it is to be used.

Modulation: The modification of one signal by another by some process other than addition. This is how audio signals are carried on radio waves (page 38).

Moment of inertia: The product of a mass and the square of its distance from a pivot. Used to assess the properties of pick-up arms on record players; the moment of inertia should be low (page 30).

Mono: A single-channel sound system producing no impression of width but simply of sound coming from a point. A mono signal cannot be made into a stereo one since it does not contain any positional information.

Muting: Simply the turning off of signal. Sometimes fitted to FM tuners to turn off automatically the noise between stations (page 40).

Negative feedback: *See* Feedback.

Glossary

Noise: Random electrical signals that exist in all circuits. Noise usually sounds like a quiet hiss. It cannot be removed entirely since it is produced by anything electrical, but its effects can be minimised by good engineering design. The signal-to-noise ratio of an amplifier is a mark of how well this has been done and should be a large figure in decibels.

Noise reduction systems: Electronic systems for reducing the effect of noise introduced in recording and record-making processes by artificially maintaining a high signal level at the recording stage and reversing the process during playback (page 84).

Ohm: The basic electrical measurement of resistance, abbreviation Ω. 1000 ohms = 1 kilohm (kΩ).

Ohm's Law: The law relating voltage (V), current (I) and resistance (R) as $V/R = I$. If you consider voltage as a force, then this boils down to "The harder you push, the more you get through if the resistance stays the same."

Open circuit: A break in a circuit so that there is no complete path for the current to take.

Open reel, Open spool: A method of using recording tape by winding it off one reel through the tape machine and on to another reel. Although inconvenient for storage, this is the method that allows easy editing.

Oscillation: Continuously changing voltage or current usually with a defined waveform and frequency (though perhaps not intentional). Considered over an integral number of cycles an oscillation averages to zero overall change, *i.e.* there is no displacement.

PPM: Peak programme meter. A meter with a logarithmic response; *i.e.* it reads on a decibel scale, which registers the peak values of signals presented to it. It has a very fast rise time and a slow, steady fall time, enabling fast transient signals to be read.

Parallel tracking: Descriptive of a pick-up arm on a record player which copies exactly the path taken by the cutting head

Glossary

used to make a record; *i.e.* either the cartridge follows a radius or the stylus is caused to remain tangential to the record groove in all playing positions (page 28).

Parasitic instability: Instability in an amplifier which occurs only at a given signal voltage displacement. Therefore the amplifier is stable under zero signal conditions. This instability causes a buzzing noise on top of programme material.

Parasitic speaker: A treble speaker built into a bass drive unit and using the same voice coil as the bass unit. This speaker takes over the transmission of sound when the massive bass cone becomes inefficient (page 62).

Phase: The timing relationship of two signals of identical frequency to each other (page 100).

Phase linear: Descriptive of amplifiers. loudspeakers, etc., when the phase change through the equipment is the same for all frequencies in the audio range (page 66).

Phasing: (*a*) Of loudspeakers: making sure that the same signal fed to two loudspeakers in the same system causes the cones to move in the same direction. (*b*) Of pop music: a special effect caused by moving the frequency of a particular type of filter.

Phono: A type of connector for coaxial leads. Probably the most common on domestic equipment. *See* Fig. 44 for wiring

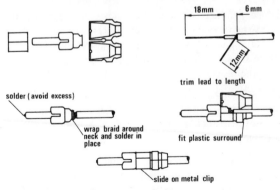

FIG. 44. Wiring of one type of phono connector

Glossary

details of one type. Also used of gramophone inputs on some equipment.

Pick-up arm: The arm on a record player which moves across the record and carries the cartridge (page 28).

Piezo electricity: Electricity generated by bending crystals of certain salts. The signal generated by a crystal or ceramic cartridge (page 23).

Pilot tone: Oscillation of known frequency and amplitude used to test or set up audio equipment.

Pinch wheel: A wheel usually covered in rubber which holds the tape against the capstan in a tape recorder during record and playback (page 79).

Pitch: Either frequency or, of musical instruments, the relationship of their basic scale to the international scale based on "A" being 440 Hz.

Playback head: The tape head in a tape recorder which takes the magnetic signal off the tape and turns it into an electrical signal for amplification (page 74).

Playing weight: The effective weight with which the stylus on a record player presses down on to the record (page 27).

Polar response: The sensitivity of a microphone to sounds coming to it from different directions (page 108).

Positive feedback: *See* Feedback.

Power: The product of the voltage across a circuit and the current flowing through it; *i.e.* where W is the power in watts, V is the voltage and I is the current in amps, $W = V \times I$. From Ohm's Law this may also be seen to be $W = V^2/R$ where R is the resistance in ohms. The latter formula is sometimes more useful.

Power amplifier: An amplifier which takes a low-level signal and increases it to such a level that it is able to drive a loudspeaker (page 54).

Preamplifier: An amplifier which normalises different input signals and allows tonal changes, etc., to be made to them (page 48).

Glossary

Pre-emphasis: Purposefully introduced changes in the frequency content of a signal to overcome deficiencies in following apparatus (page 42).

Presence: Slight exaggeration of upper mid frequencies in an audio signal to give a slightly more "lively" sound to the programme material (page 52).

Programme: Audio signal designed to be listened to, rather than test signals, etc.

Q factor: A description of how narrow a range of signals a filter designed to pass one centre frequency actually passes. Where Q = the Q factor, f_0 = the centre frequency and f_B = the bandwidth, $Q = f_0/f_B$.

Quadrophony: A method of using four loudspeakers each fed with different but related signal to reproduce an audio environment surrounding the listener.

RF: Radio frequency. Generally descriptive of the radio frequency a receiver is tuned to, but actually any frequency above 30 kHz.

RIAA: The pre-emphasis curve used in the manufacture of records. RIAA equalisation is the de-emphasis used in a preamplifier to recover the original frequency content of the signal (page 49).

Radio: (*a*) The transmission of signals by coding them on to electromagnetic radiation. (*b*) A piece of apparatus for receiving such signals (*see* Chapter 4).

Record head: The tape head on a tape machine that forces the audio signal on to the tape as a magnetic signal (page 73).

Reel: The plastic or metal drum on which recording tape is stored on an open-reel tape recorder (page 70).

Reel to reel: Another term for an open-reel tape recorder.

Reflex: A type of bass loudspeaker enclosure in which there is a controlled sound path from the back to the front of the drive unit (page 63).

Glossary

Resistance: Exactly what it sounds like – the resistance an electronic circuit puts up to the passage of electricity. Measured in ohms (Ω).

Response: The way the output of a piece of apparatus behaves for changes in one of the parameters of its input signal.

Resonance: A tendency for a piece of equipment or a room to oscillate more easily at one frequency than at others, so exaggerating that frequency.

Reverberation: The property of sound of not stopping immediately the source is stopped but dying away slowly through multiple reflections off the listener's surroundings. Sometimes added artificially to recordings.

Rumble: A low-frequency noise in record players often caused by the mechanical rotation of bearings transmitting vibrations which are picked up by the stylus.

SW: Short wave. The highest frequency band in general use for AM radio transmissions (page 38).

Screened cable: A cable in which the conductor or conductors have an overall earthed sheath of conducting metal or plastic insulated from them but preventing outside signals from interfering with them. Usually used for connecting low-level sources to sensitive inputs.

Screening: The technique of putting an earthed metal box round a sensitive circuit to prevent interference from outside. Also any earthed piece of metal used to prevent two circuits interacting. (*See* Screened cable.)

Sensitivity: The input voltage required to drive an amplifier to maximum output.

Short circuit: Zero resistance.

Sine wave: The simplest waveform, from combinations of which all others may be made. It sounds very boring but is invaluable for test purposes (page 99).

Single play tape, Standard play tape: The thickest variety of recording tape. It is the most robust, but the thickness restricts the length available on a spool.

Glossary

Signal-to-noise ratio: (*a*) For power amplifiers: the ratio of the full power output signal to the noise output with zero signal. (*b*) For other equipment: the ratio of the normal output signal to the noise output with zero signal. These ratios are normally measured in decibels.

Signal strength meter: A meter on a radio tuner giving a reading proportional to the radio signal being received.

Solid state: Used of electronic components which rely for their operation on the physical properties of solids, as opposed to valves, where the physical properties of a vacuum are used. Solid-state circuitry is more rugged than valve circuitry.

Sound: Pressure oscillations in the air lying between 20 Hz and 20 kHz in frequency (*see* Chapter 10).

Speakers: Abbreviation for loudspeakers.

Spherical stylus: A record player stylus the tip width of which is the same as its breadth, both being about 0.0005 in. (page 26).

Squelch: A circuit which cuts off the output of a tuner when the RF signal strength goes so low as to give unacceptable audio signal quality.

Stereo: A method of using two loudspeakers each fed with a different but related signal to give the impression of a sound source spread out in front of the listener.

Stylus: The small point at the tip of the cartridge on a record player, usually of diamond or sapphire, which follows the groove of a record.

Tape: A thin film of plastic coated with a magnetic oxide used in a tape recorder to carry the recording as a magnetic signal (page 82).

Tape head: Any of the transducers in a tape machine used to erase unwanted signals, record new ones or play back recorded ones (page 72).

Tape player: A tape machine that can play back pre-recorded tapes but cannot record tapes itself.

Glossary

Tape recorder: A machine that can record signals on to magnetically coated tape and replay them when required, also having the facility to erase previous recordings if tape is to be reused (*see* Chapter 7).

Test tone: A continuous sine wave used for testing.

Tone: The quality that a chosen frequency response imparts to programme material.

Tone controls: The controls on a preamplifier that adjust its frequency response (*see* Chapter 5).

Track: Either one of the signal-carrying strips on a recording tape (page 75) or one section of programme material on a record, though not necessarily one whole side.

Transducer: A piece of equipment that changes electrical signals into some other form (sound, pressure, displacement, etc.) or vice versa.

Transformer: A piece of apparatus that changes the voltage and current amplitudes of a signal without altering its power. Transformers are used among other things to produce low voltage from mains efficiently, in order to provide power supplies suited to transistors. A transformer will operate only with a.c.

Transistor: An electronic component which allows a small current to control the value of a larger one. Used as the basis of most modern amplifiers.

Transmission line speaker: A loudspeaker in which the back of the drive unit has transmission line loading (page 64).

Transmitter: Radio frequency circuitry which drives an aerial to send electromagnetic radiation over a distance. Alternatively the equipment which transmits a radio signal.

Treble: The upper portion of the audio-frequency spectrum.

Triple play tape: Recording tape made one-third the standard thickness enabling three times as much to be stored on a given size of reel. It is extremely fragile and suffers from print-through, but storage is more convenient.

Glossary

Tuner: A radio receiver, especially for FM transmissions. It may or may not have an amplifier incorporated to drive loudspeakers (Chapter 4).

Tuning: The action of selecting a radio station on either AM or FM receivers.

Tuning meter: A meter that indicates how well a radio receiver is tuned, usually centre zeroing with deflection at left or right to indicate which way the tuning knob should be turned for better reception. The function of this meter may be performed by a pair of lights.

Turntable: The round disc on a record player which carries the record and rotates at constant speed (page 32).

Tweeter: A loudspeaker drive unit designed to carry treble signals only (page 58).

UHF: Ultra-high frequency. The radio frequency band on which 625-line television is transmitted.

VHF: Very high frequency. The radio frequency band on which FM radio is transmitted; also used for 405-line television transmissions.

VU meter: Volume unit meter. A cheaper form of meter for showing audio signal amplitudes. It is not logarithmic and so does not really indicate how loud a signal is; it also has a slow response so that fast transient signals are ignored. This type of meter is usually fitted to tape machines as a recording level indicator.

Valve: An electronic component which allows a small voltage to control the value of a larger one. Used as the basis of older amplifiers and some modern ones where high voltages have to be controlled.

Volt: The unit of force in electricity. A millivolt is one-thousandth of a volt, so 1000 mV = 1 V. (*See* Ohm's Law.)

Volume: The overall perceived amplitude of a sound or of programme material.

Glossary

Watt: The unit of power in electricity, it is the product of voltage and current; so where V is the voltage across a circuit and I is the current flowing, then $V \times I = W$.

Woofer: An old term for a loudspeaker designed to carry bass signals only (page 85).

Wow: A slow fluctuation in pitch caused by a record player or tape recorder not running at constant speed.

NOTES

NOTES